"There is no better place to start learning a
Törneke teaches the principles of RFT sim
of clinical examples to make it accessible a
this had existed when I first learned about RFT; it would have saved me many hours of hard work, frustration, and confusion."

—Russ Harris, author of *The Happiness Trap* and *ACT Made Simple*

"On rare occasions, the skills of writer, therapist, and theorist combine to give the field a sophisticated yet highly practical book. This much-awaited translation shows relational frame theory as an accessible, powerful tool for all who use talk therapy. A must-read for those interested in contemporary behaviorism."

—Kelly Koerner, Ph.D., creative director at Evidence Based Practice Institute in Seattle, WA

"Since RFT first appeared in the experimental literature, it has been hailed as a breakthrough in our scientific understanding of language and cognition with direct and important implications for clinical psychological practice. Yet, descriptions of RFT, written largely for technical audiences, have been, at best, curiously baffling, and at worst, maddeningly incomprehensible. In this book, Törneke has solved the puzzle of RFT! He summarizes the history of RFT, its key features, and its clinical implications with language that is user-friendly and easily understandable. I believe this book will make a huge difference for clinicians who wish to understand RFT and its implications for clinical practice. It also may be a useful learning tool for researchers and RFT experts themselves who wish to learn and see a beautiful example of how RFT can be presented clearly and comprehensively."

—Jonathan Kanter, associate professor at the University of Wisconsin-Milwaukee and director of its Depression Treatment Specialty Clinic

"For years, clinicians have asked me for recommendations about what they should read to learn RFT. There was really no good advice I could give except 'be persistent.' Finally, I have a better answer. If you want to understand relational frame theory, this is the place to start. Törneke's RFT primer is both masterful and accessible."

—Kelly G. Wilson, Ph.D., associate professor of psychology at the University of Mississippi, coauthor of *Acceptance and Commitment Therapy* and author of *Mindfulness for Two*

"At times, while reading Törneke's book, I have felt as though I were in the middle of a thriller about the psychopathological behaviors of humans. Clues to unraveling the mystery embedded in complex concepts like 'arbitrarily applicable relational responding' have alerted me, as the reader, to what is coming up next. Our ability for relational framing and for rule-governed behavior may at first glance seem fabulous—a gift from the gods—but darkness lurks around the corner. Our ability to problem-solve is the villain. This book helps me make sense of it all."

—Maria Midbøe, M.Sc., candidate in psychology at Stockholm University in Stockholm, Sweden

"Until now, explanations of relational frame theory have remained largely esoteric and even impenetrable to all but the most specialized scholars. For the first time, this extraordinary book provides a highly accessible account of relational frame theory, including its larger context within psychology, the current research in the field, and its many potential applications. Törneke strikes a fine balance between doing justice to relational frame theory and making the theory, research, and its implications readily comprehensible to the non-expert. This unique book is a must-read for scholars of human cognition, as well as clinicians, educators, others seeking to harness the power of basic psychological principles in their applied work, and anyone interested in the renaissance of modern behavior analysis."

—James D. Herbert, Ph.D., professor of psychology at Drexel University in Philadelphia, PA, and director of Anxiety Treatment and Research

Learning RFT

An Introduction to Relational Frame Theory and Its Clinical Application

NIKLAS TÖRNEKE, MD

CONTEXT PRESS
An Imprint of New Harbinger Publications, Inc.

Publisher's Note

This publication is designed to provide accurate and authoritative information in regard to the subject matter covered. It is sold with the understanding that the publisher is not engaged in rendering psychological, financial, legal, or other professional services. If expert assistance or counseling is needed, the services of a competent professional should be sought.

Distributed in Canada by Raincoast Books

New Harbinger Publications, Inc.
5674 Shattuck Avenue
Oakland, CA 94609
www.newharbinger.com

Printed in the United States of America

Acquired by Catharine Sutker; Cover design by Amy Shoup; Edited by Jasmine Star

---

Library of Congress Cataloging-in-Publication Data

Törneke, Niklas.
[Relational frame theory. English]
Learning RFT : an introduction to relational frame theory and its clinical applications / Niklas Törneke ; foreword by Dermot Barnes-Holmes ; afterword by Steven C. Hayes.
p. cm.
Includes bibliographical references.
ISBN 978-1-57224-906-6
1. Cognitive therapy. 2. Acceptance and commitment therapy. I. Title.
RC489.C6313T67 2010
616.89'1425--dc22

2010024048

14 13 10 9 8 7 6 5 4 3

In remembrance of my father, David

Contents

Clarity, simplicity, depth.
This is what I seek to achieve and unite.
Excluding none of the three.
Hence the difficulty.

—Pär Lagerkvist (translated by Elizabeth Ask de Lambert)

Foreword to the U.S. Edition

A PRAGMATIC THEORY OF HUMAN LANGUAGE AND COGNITION

Behavior analysis is an extremely unusual approach to psychological science. In stark contrast to mainstream psychology, the behavioral tradition refuses to appeal to mediating mental representations and processes as the basis for explaining human behavior. Instead, it adheres to a thoroughgoing functional analytic approach in which the systematic analysis of the interactions between an organism and its past and current environmental contexts provides the framework for explaining all psychological events.

Somewhat surprisingly, perhaps, this unusual approach yielded considerable success in the early years, particularly in improving the lives of individuals diagnosed with various learning disabilities. However, the same level of success was not observed when behavior analysis turned its attention to human language and cognition. Noam Chomsky's damning review of B. F. Skinner's *Verbal Behavior* is well known, and is sometimes offered as "proof" that the nonmediational approach of behavior analysis could not stretch to the more advanced or sophisticated aspects of human psychology (such as language and thought). In fact, Skinner's work did go on to provide the basis for a number of language training programs, but once again success was limited largely to learning-disabled populations.

The key problem with *Verbal Behavior*, and one that Chomsky highlighted, is that it fails to address the highly generative nature of human language. Although the book does not leave this issue completely untouched, it fails to provide a well-developed technical account of the almost infinite novelty that language can generate. Furthermore, its treatment of advanced language phenomena such as metaphor and analogy is unconvincing. For example, in providing functional analytic interpretations of these verbal behaviors, it interweaves lay terms with technical terms, and the resulting analyses thus lack the required precision.

But of course Skinner wrote the book almost two decades before Murray Sidman's first study on equivalence class formation was conducted, with all of the implications for the behavioral study of human language that emerged from that seminal research. Skinner was clearly at a serious disadvantage in not having access to this equivalence data set and the conceptual work that followed. We now know that derived relational responding appears quite early in the behavioral repertoires of young children, and modern behavioral treatments of human language and cognition have emerged from this work. Unfortunately for Skinner's *Verbal Behavior*, this research was lacking in the book. For the most part, therefore, *Verbal Behavior* was a direct contingency account of human language that made only passing reference to the most important defining feature of verbal behavior: derived relational responding.

The first book on relational frame theory, *Relational Frame Theory: A Post-Skinnerian Account of Human Language and Cognition* (Hayes, Barnes-Holmes, & Roche, 2001), published over thirty years after Skinner's work, aimed to present a modern behavior analytic account of human language and cognition. The theory embraced derived relations and indeed put them at the very heart of the account. Nevertheless, RFT remained a natural extension of earlier conceptual and empirical research within behavior analysis. The core concept in the book, arbitrarily applicable relational responding, was based solidly on Skinner's concept of the operant and drew heavily on Sidman's seminal work on equivalence classes. Specifically, equivalence class formation was seen as the result of a history of operant conditioning (a learned response class), and based on this argument, the possibility of multiple forms of such response classes (relational frames) was predicted.

The 2001 RFT book explains how the basic analytic units of human verbal behavior—relational frames—may be combined into more complex units, giving rise to the relating of relational frames, increasingly complex relational networks, and the relating of entire relational networks to other relational networks. These types of highly abstract concepts are used in the book to provide a nonmediational and purely functional analytic account of the full range of human verbal abilities, including naming, storytelling, humor, abstract logic, the verbal construction of self, and spirituality.

The primary purpose behind the 2001 RFT book involved providing more than a modern behavioral *interpretation* of human language and cognition. Its purpose was intensely pragmatic. Among other objectives, the book aimed to stimulate both basic and applied research on human language and cognition, and to provide a set of functional analytic terms that would facilitate communication among researchers and practitioners. It appears to have been relatively successful in achieving the first goal, but the latter objective, I believe, requires another book: the one you are currently reading.

Relational Frame Theory: A Post-Skinnerian Account of Human Language and Cognition is intensely academic, full of jargon, and littered with highly abstract concepts. *Learning RTF* contains many of those concepts and some of the jargon, but it presents the material in a very accessible manner and, critically, does complete justice to the subject matter.

The first section of *Learning RFT* starts with a succinct but well-worked introduction to the philosophical and conceptual underpinnings of behavior analysis, an understanding of which is essential in grappling with what is to follow. The topics of thinking and human language are then introduced, and the traditional Skinnerian perspective on these topics is explained and contrasted with that of traditional cognitive therapy. In this examination it is proposed that neither approach has fully dealt with the role of thinking and language, at least in the clinical domain. The first section of the book will serve as a strong motivator for the reader, and particularly the clinician, to delve into the next and perhaps most challenging section.

In the second part of *Learning RFT*, the theory itself is presented, but in a highly accessible way. The chapters in this section strike a perfect balance between providing an appropriate level of technical detail and keeping the writing lively, light, and a pleasure to read. Furthermore, although the earlier chapters in this section focus necessarily on the more abstract features of RFT, the writing progresses rapidly and with relative ease to issues that will be of more interest to the practicing clinician, dealing with topics such as self and perspective taking. The final chapter in this section, "The Dark Side of Human Languaging," will be particularly relevant to clinicians in that it explains how RFT means that human language and cognition may be the source of much human suffering.

The third and final section of the book focuses on the clinical implications of RFT. The section begins with traditional behavior therapy and how it relates to other therapeutic approaches, then it explains how RFT makes a unique contribution to our understanding of psychotherapy itself. The reader is now prepared for the final chapters of the book, which work systematically through the application of modern behavior analysis to clinical psychology. This material provides a powerful review of clinical behavior analysis and in particular explains how RFT supplements and extends the traditional behavior therapeutic approach. It is only in this final part of the book that the intensely pragmatic nature of RFT is fully revealed. This highly abstract and arcane theory allows the practitioner to conceptualize human language and thought as composed of behavioral units that may be subjected to functional analyses and behavioral intervention strategies. In short, *Learning RFT* clearly illustrates in a very powerful way how RFT can contribute toward the conceptualization and treatment of human suffering. In truth, this is a book I would love to have written.

—Dermot Barnes-Holmes
National University of Ireland, Maynooth

Dermot Barnes-Holmes is foundation Professor of Psychology at the National University of Ireland, Maynooth. He has published approximately two hundred scientific articles, book chapters, and books, the vast majority of which have focused on the study of human language and cognition from a behavior analytic perspective.

Foreword to the Swedish Edition

Is there a need for a book devoted to relational frame theory? Can we learn anything about language by endorsing the principles of learning? Honestly, is it not the case that cognitive psychologists do a much better job at explaining how we think? My answers to the first two questions are affirmative, and *Learning RFT* is an important contribution, giving a thorough description of how a behaviorist framework can help us understand cognition and language. While some readers might not find this topic attractive at first sight, perhaps because they have a clinical focus, my conviction is that this book can be helpful for clinicians as well as researchers and that the principles described are put forward in a reader-friendly manner, facilitating comprehension of the sometimes difficult-to-grasp concepts in relational frame theory.

I received my training as a clinical psychologist in the mid-1980s, and most of the textbooks I read stated that behaviorism was dead and that the cognitive revolution had taken over after the "dark years" under the rule of B. F. Skinner. However, not all my teachers shared that opinion, and at the department of psychology in Uppsala, Sweden, I got the basics of applied behavior analysis and developed an interest in behaviorism. I guess I can say I acquired what might best be called a "behaviorist framework," which influenced my clinical work and research, by that time focused on hearing loss in the elderly. In my work I found operant psychology very useful and ended up with a thesis entitled *Hearing as Behavior*. However, I was also painfully aware of the fairly low status of operant psychology in mainstream psychology, and while I found the work by Skinner useful, I could not fully appreciate his book *Verbal Behavior*. On the other hand, the work by Steven Hayes and his colleagues on rule-governed behavior was in my awareness, but I never saw it mentioned in the cognitive psychology literature. While I could understand the objections to Skinner's analysis of language, I thought Chomsky's review from 1959 was overly negativistic. All of that said, the psychology of learning and behavior still makes a difference, and the development of relational frame theory (RFT) clearly indicates that it was premature to dismiss behaviorism as a framework for the understanding of language and cognition. When many researchers and students of psychology believed behaviorism was long gone, interesting activities continued to occur in this field. *Learning RFT* provides an excellent summary of what happened after Skinner, and also does justice to what Skinner probably meant with his analysis of verbal behavior.

What does RFT add, then? I am not too convinced that RFT is necessary for how we provide effective psychological treatments, but I do believe we can benefit from a good theory, and in addition to wanting to be good clinicians, we also want to be able to understand and explain human behavior. Since language and thinking are integral parts of how we understand each other in lay terms, it makes sense that behavior therapists were attracted to cognitive therapy, in which thoughts and beliefs were prioritized. However, not all behavior therapists were convinced. In clinical psychology, and in particular in the field of psychotherapy, it could be argued that basic psychological science, such as cognitive psychology, is used to inform clinical practice while not really being part of clinical science. For example, clinicians sometimes regard Aaron Beck's cognitive therapy as part of cognitive psychology, but in reality, basic cognitive psychology on constructs such as working memory has only more recently been applied in clinical research within the psychotherapy field. Most of what has been written about cognitive therapy has very little in common with basic cognitive psychology. Within the behavior therapy community, this has sometimes led to a negative attitude toward what cognitive psychology can contribute, as we sometimes falsely assume that cognitive therapy equals cognitive psychology. This might be a mistake, as overviews of the scientific status of different subfields in psychology clearly show that cognitive psychology is the leading branch of psychology when it comes to research funding, publications, and citations. But perhaps the emergence of RFT will change the situation for behavior therapists with an interest in language and cognition. It is probably fair to say that behaviorists had not finished their work on language and cognition in the lab, and I hope the work on RFT will inspire therapists to look more closely at basic research in language and cognition. Even if knowledge of RFT isn't necessary to do a good job as a clinician, good explanations of human behavior definitely are. Also, because of our background in psychology, we need data to be convinced that something is likely to be true. With RFT, we now have more tools to explain the verbal behavior we confront in our clinical work.

You may wonder about the third question I raised: Is it not true that cognitive psychologists are doing a better job at explaining cognition? *Learning RFT* is excellent in this respect, as the author does not ignore the substantial literature on the cognitive psychology of language; rather, he relates it to RFT. For me, and most likely for many other psychologists, this makes it easier to understand RFT and take it seriously as a major contribution to psychology. So here's the answer to my last question: To date, cognitive psychology is, if not the best, at least the most productive when it comes to language and cognition. But RFT need not be seen as in opposition to the rest of psychology, and it can provide us with important clues to further our understanding of language and cognition. It is possible that a behaviorist renaissance is on the horizon. Niklas Törneke's book is one of the building blocks in that venture.

—Gerhard Andersson, Ph.D., professor of clinical psychology in the Department of Behavioral Sciences and Learning at Linköping University, in Linköping, Sweden

Acknowledgments

In 1998 I went to an international conference in Ireland and for the first time heard two people speak who, more than any others, are behind the ideas and the research this book is based on: Steven Hayes and Dermot Barnes-Holmes. From that time, both of them have generously answered my questions and helped me become familiar with an outlook and research tradition that, up until then, had been essentially unknown to me. Many thanks to both!

The person who first told me I should write a book like this is Kelly Wilson. He is also the person I have mainly learned ACT from in practice. I owe him warm thanks, as well.

A fourth person who has meant a lot to me in the process leading to this book is Carmen Luciano. She too is a leading figure in the international network of researchers and clinicians bound together by a common interest in RFT and ACT. In recent years she has been a never-ending source of knowledge and inspiration to me.

Several individuals in Sweden have also been of particular help to me. First and foremost, Jonas Ramnerö. Ever since we met at a conference in Dresden in 1999, we have been engaged in an ongoing dialogue on the role of behavioral psychology in psychotherapy. This dialogue has made a decisive contribution to my writing this book. Jonas has also helped me by reading and commenting on the Swedish manuscript, as have Jonas Bjärehed, Martin Cernvall, and Billy Larsson.

As for the English version, Kelly Koerner, Rainer Sonntag, and Ian Stewart have all read parts of an earlier version of the manuscript and made many valuable suggestions. All deficiencies remain, of course, my own responsibility. Elizabeth Ask de Lambert did most of the work translating the book into English, and Jasmine Star made the editing process extremely helpful and smooth. Görel Gunnarsson and her colleagues in the medical library at the county hospital in Kalmar, Sweden, have been of invaluable help in obtaining articles and other literature. My heartfelt and sincerest thanks to one and all!

A Personal Word of Introduction

As a psychotherapist, I am a child of my time. I grew up in a psychodynamic world dominated by a strong emphasis on understanding, but not—as far as I could see—well anchored in scientific research. It also lacked concrete guidance regarding workable therapeutic interventions. My encounter with cognitive therapy at the end of the 1980s was therefore a liberating experience. Scientific foundation was paramount, and the therapeutic strategies were applicable in my everyday work within psychiatry. Cognitive theory as a basis of psychotherapy has since been victorious—not only in my world, but in the world of psychotherapy at large. Cognitive therapy has gradually been integrated with behavior therapy under the designation of "cognitive behavioral therapies" (CBT), but the different hybrids are dominated by the theories underlying the cognitive model.

During the 1990s, I progressively found what I saw as shortcomings in cognitive theory. It was difficult to get a clear idea of what the basic terminology was and what scientific support it had. Different cognitive theorists liberally used their different terms to describe what was taking place in the "psyche," and this obscurity and lack of consensus became all the more troublesome due to the assumption that the core of psychological problems was located somewhere inside this unknown realm. In the mid-1990s, I was still largely unaware of any vigorous alternative. I did, however, run into some works that aroused my interest, notably Marsha Linehan's 1993 book on dialectical behavior therapy (DBT), which revealed a strong influence from classical behavior therapy.

Through my interest in affect theory and the use of metaphors, I came into contact with acceptance and commitment therapy (ACT), a therapy model related to DBT in many respects, but with a much more elaborate theoretical and experimental foundation. Once again I was provided with new, useful tools. In addition, I found myself in more extensive contact with the new theoretical approach ACT had been built on, as well as with classical learning theory as a basis for psychological therapy. I realized that one cannot really understand ACT, much less its theoretical foundation, relational frame theory (RFT), without an understanding of basic behavioral principles—that is, operant and respondent conditioning. Once I gradually became familiar with these principles, I discovered what I have come to regard as the most promising psychological model available when it comes to understanding human behavior in a way that also contributes directly to the therapeutic work for change. This led to an inspiring dialogue with my good friend Jonas Ramnerö, who'd had a somewhat longer history with behavioral psychology than I had. This dialogue resulted in our writing a book together: *The ABCs of Human Behavior: Behavioral Principles for the Practicing Clinician* (Ramnerö & Törneke, 2008).

Recent years have seen a growing interest in behavioral principles, not least among psychotherapists with cognitive training. This new interest has given me a reason to work more thoroughly on the issue that has engaged me for quite a few years by now: the issue of the power of thinking in relation to other human behavior and the role this plays in the problems that make people seek help in psychotherapy. My purpose in this book is to elucidate the behavioral perspective on this question. You might say that this book attempts to respond to the questions raised by cognitive theory and therapy, although the answers are given from a different point of departure than those common to cognitive approaches. Instead, the starting point here is the foundation laid down years ago in the shape of operant and respondent conditioning. Behavioral psychology has long had problems in applying its basic agenda of prediction and influence in this area. I think these problems are coming to an end. The increasing basic research performed to describe relational conditioning and the theoretical structure that has evolved around this phenomenon (RFT) provides new answers and opens the door to new interventions in respect to cognition and human language. All this is done from the agenda of classical behavioral therapy, with new interventions being founded upon data from experimental research.

For me to begin this book by contrasting its content with psychodynamic and cognitive models may seem provocative and deprecating. This is not my intention. I am aware that the psychodynamic and cognitive theories that I earlier found unsatisfactory have not stayed unchanged since my point of departure; they have been developing in their own way. Moreover, I believe that a behavioral perspective is integrative in nature. Behavioral psychology is not about one specific model of therapy; it describes fundamental, universal principles of behavior. It is legitimate to approach anything and everything a person does from this perspective. Behavioral psychology is not limited in scope to phenomena that are easy to observe and define, like when someone avoids air travel or washes her hands obsessively. It also applies to behaviors in the close interactions between people that are harder to capture, like establishing and maintaining close relationships or behaving toward the therapist much like one earlier behaved toward a parent. Even behavior that in many ways seems concealed to anyone other than the person who is doing it, like dwelling on past grudges or struggling with feelings of dejection, can be approached from this perspective. The same goes for behaviors that are rarely the focus in psychotherapy, like playing the flute or writing poetry.

Thus, nothing human is alien to behavioral psychology. All human phenomena that take place in time and space and that can be the subject of attention in psychodynamic or cognitive therapy can also be approached from this perspective, hence its integrative nature. But the starting point of this analysis is a given theoretical position: the one which, since the days of Skinner, has been designated "radical behaviorism." Some of this is novel, but the foundation was laid quite a few years back.

A WORD ON TERMINOLOGY

Skinner called the science of behavior that he developed *behavior analysis*. However, this term is used in slightly different ways. Within behavior therapy in Europe, it is sometimes used synonymously with the word "conceptualization." In this usage, a behavior analysis is understood as an initial phase of behavior therapy. I will be using the term in the way Skinner did, which is how it is still used in the United States. Used in this way, "behavior analysis" is a designation of the science, as a whole, that aims at predicting and influencing behavior, along with the practical work involved in doing this. There is usually a distinction made between two branches within behavior analysis: experimental behavior analysis and applied behavior analysis. *Experimental analysis of behavior* is the type of experimental activity usually connected with Skinner: Under carefully monitored conditions, different factors are varied to determine whether an organism's behavior can be predicted and influenced. In *applied behavior analysis*, the basic principles that can be described following the experimental work are applied to different types of problems "out there, in real life." A branch of applied behavior analysis is *clinical behavior analysis*. This is behavior analysis in the field commonly referred to as psychotherapy. Consequently, in our book *The ABCs of Human Behavior: Behavioral Principles for the Practicing Clinician* (2008), Jonas Ramnerö and I called this field of application "behavioral psychotherapy." In this book, however, I will mainly be using the term "clinical behavior analysis."

OUTLINE

This book is divided into three parts. Part 1 provides some important background. Chapter 1 offers a short account of basic and well-known principles of learning from the viewpoint of radical behaviorism, with a particular emphasis on concepts that must be understood in order to become familiar with RFT. Chapter 2 provides a survey of how behavior analysis had tried to tackle "the power of thinking" before the experimental data on which RFT is based were available. The bulk of this chapter consists of an overview of Skinner's analysis of verbal behavior. Although his analysis has limitations (described here as well), it remains important as a backdrop to RFT. In chapter 3, arguments for renewed inquiry into human cognition and language conclude part 1 of the book.

Part 2 of the book is its core; this is where RFT is described. Chapter 4 presents and defines RFT's basic terminology and describes the type of experiments the theory is based on. In essence, chapter 4 describes the fundamental elements in human language. In chapters 5 and 6, I have attempted to show how these building blocks are combined with an increasing degree of complexity, and how they cast new light on complex human behavior. In chapter 7, part 2 concludes with an account of the problems that verbal (cognitive) behavior creates for human beings, or the side effects of human language.

Part 3 of the book describes clinical applications. Chapter 8 takes a look at psychological therapies in general from a behavioral perspective. The remaining three chapters focus on clinical behavior analysis, with particular emphasis on strategies and techniques based in RFT.

THE CHARACTER OF THE BOOK

The number of scientific texts presenting RFT and its experimental base is growing rapidly. The same is true for books presenting ACT. This book has the character of being in between these two categories of texts. Although the purpose is to give an overall introduction to RFT, the book has its limitations. The main limitation lies in maintaining both a theoretical and a clinical perspective. Though RFT is based on experimental research, this book does not present the experimental work in detail; it simply gives an outline of the experimental work and devotes more attention to the conclusions drawn from that work. This book is more focused on concepts than data and details, partly to give a general introduction, and partly to give an understanding that facilitates clinical work. The book does not include more detailed presentations of the experiments in their entirety, such as how they are arranged and performed. I have tried, however, to frequently refer to literature that contains such presentations so that the interested reader can find more in-depth material. There is also a paradox involved in this limitation. There is a degree of learning RFT that can be achieved only by engaging in experimental work. Yet this is a book by someone who has never done that, written primarily for others in the same situation. Experimental psychologists will probably find lack of precision and technical detail. The same might be true of others who are very well acquainted with the existing scientific literature. At the same time, some readers will probably find parts of the book too technical and abstract. Still, this kind of book, "in between," is what I wished to read when I first encountered RFT. Hopefully it will be helpful to others who are now in the situation I was in then.

ACT has a central position in part 3 of the book, on clinical applications. This is only natural, as this therapeutic model has evolved together with RFT. Alongside ACT, other forms of clinical behavior analysis, especially functional analytical psychotherapy and behavioral activation, have their place. It has not been my goal, however, to present any of these individual models in their entirety, or to carry out an in-depth comparison. I want to pursue the agenda outlined in *The ABCs of Human Behavior* (Ramnerö & Törneke, 2008): to describe psychological therapy from the broad perspective of radical behaviorism, and to describe the therapeutic tradition that can be called behavior therapy, behavioral psychotherapy, or clinical behavior analysis. I want to do this with a special emphasis on how an understanding of RFT adds some new elements to this tradition.

PART 1

Background

CHAPTER 1

Radical Behaviorism and Fundamental Behavior Analytic Principles

Radical behaviorism is the philosophical basis on which B. F. Skinner built his psychology (1953). This term has caused a lot of debate over the years. It has also frequently been misunderstood, sometimes to such a degree that one might wonder if the term is useful at all or if it has actually become an obstacle to introducing the psychology itself. It is common to see Skinner's views described as superficial and coarsely mechanistic, even in psychology textbooks (Power & Dalgleish, 1997, pp. 35-36; Solso, MacLin, & MacLin, 2005, p. 329). This is in stark contrast to my own impression from reading Skinner, and I have often wondered if writers who portray his positions in this way have actually read his works. Be that as it may, his positions, and the terms he uses to describe them, are controversial, and the term "radical behaviorism" is a clear example of that. An alternative and more modern term is *functional contextualism* (Gifford & Hayes, 1999). This alternative term may better convey in what way this particular philosophy of science relates to other modern approaches. It puts Skinner's position in relation to alternative types of contextualism, such as social constructivism or certain types of feminism (Roche & Barnes-Holmes, 2003; Gifford & Hayes, 1999). The term "functional contextualism" emphasizes two essential elements in radical behaviorism: The first is that behavior must always be understood in relation to the setting, or context, in which it takes place. The second is that in order to understand and influence behavior, we need to study its function—that is, what it is aimed at.

My reason for using the term "radical behaviorism" to start out with, despite the above discussion, is that this is the term that has survived and is widely accepted among those who follow in Skinner's footsteps in their work. It is linguistically correct, and it also brings out some essential elements of the position on which this book is based.

BEHAVIORISM AND ITS PREMISES

Let's begin with a discussion of the more general term "behaviorism." This is a broad term that encompasses many partly differing approaches (O'Donohue & Kitchener, 1998). Yet these approaches have certain premises in common, making the concept of behaviorism appropriate. The most fundamental premise was made clear by Watson, who came up with the term "behaviorism." In this approach, what is in focus is behavior (Watson, 1929); that is, something a person—or any other organism—is doing. The actions, or responses, of the whole organism are in focus. Another premise that the different approaches have in common is the method of seeking knowledge: The science is to be built from the bottom up. The quest is for fundamental, universally valid principles for understanding behavior. This means that laboratory experiments have a strong position. When carrying out experiments, a key concept is to try to minimize noncontrollable variables as much as possible before going on to identify and systematically manipulate those variables that are essential. In many ways this resembles how psychoanalysts are strict about the setting for their sessions, so as to eliminate irrelevant disturbances and observe relevant, governing phenomena in the interaction taking place. The most well-known, classical examples of this method within behaviorism are perhaps Skinner's experiments with pigeons and rats. The environment is stripped (there is only a box), and the relevant variables are few (the box contains a bar that enables the animal to act in order to receive food, and a light that is turned on and off). What is important here is not the laboratory experiments as such, or even the actions of pigeons or rats; rather, the aim in using this method is to be able to identify principles underlying the behavior of organisms—principles that can then be used to understand more complex processes that perhaps cannot be studied in laboratories.

This reveals another premise of behaviorism: that of an assumed continuity across different organisms. Research on pigeons, for example, is used to draw conclusions about human beings, at least in certain respects. This has sometimes been a point of controversy, especially among psychotherapists.

Is it possible to understand humans based on an understanding of animals? This was the topic of heated discussion during the 1960s and 1970s, but a lot of water has flowed under the bridge since then. Today, it is uncontroversial to say that evolutionary psychology and neuropsychology, as well as ethology, hold firm positions in shaping various theories within psychotherapy, regardless of the specific camp. Based on an evolutionary perspective, researchers who have greatly influenced different psychotherapy traditions, such as John Bowlby (attachment theory) and Joseph LeDoux (affect theory), take the same view as what Skinner once maintained: simply stated, that evolution continuously builds on what is already at hand. Functions that work are not removed; they become building blocks in future developments. This is why we can learn a lot about human beings by studying how gorillas relate to their offspring (attachment theory) or by studying basic cerebral functions in animals (affect theory).

Now, if behavior is what is to be studied, the question is how we define this term. What should be counted as behavior? The answer to this question can be expressed in somewhat different ways within behaviorism, and in order to account for what is meant by behavior in this book, I will now turn to Skinner's view on this, as encapsulated in the term "radical behaviorism."

WHAT IS RADICAL ABOUT RADICAL BEHAVIORISM?

Being radical can be taken as being extreme. This is not what Skinner had in mind, though, when he chose the term "radical." In this context, "radical" implies not "extreme" but "consistent." Radical behaviorism entails not a departure from fundamental behavioristic principles but the application of them in an all-inclusive way. This has some consequences. For example, take the principle that Skinner used to describe operant conditioning (more on this below). This principle implies that our actions are influenced by the consequences we have previously encountered following a particular action. The probability that a pigeon will peck at a certain point increases if it has earlier received food after pecking at that particular point. But if this is to be applied in a consistent way, in keeping with Skinner's position, then this principle also holds for me as a scientist. I do what I do (in my experiment with the pigeon) as a consequence of outcomes of similar experimentation earlier. As a scientist, I do not hold any objective or exclusive position. I am not outside or above the principles I study. If this understanding is applied consistently, all claims of representing the ontological truth have to be dropped. Based on this position, we cannot maintain that "this is the way it really is." Radical behaviorists repudiate the notion that the scientist operates from an objective and neutral position. As mentioned earlier, from the perspective of radical behaviorism you cannot understand behavior without studying its context. All behavior takes place within a context. But neither can the context be studied independent of behavior. This is because the scientist's attempt to study something is a behavior as well. After all, the object of our study is something that we are acting upon, just by studying it. So just as we cannot understand behavior without context, there is no context available for the organism without behavior.

This point about the behavior of the scientist is also true in a more general sense. Stimulus and response (behavior) are codependent and should be considered together. They make up a single unity (Kantor, 1970). We can separate them for practical reasons, with a certain aim in mind. And the behavioral science that Skinner wanted to create has an aim: to predict and influence behavior. Radical behaviorists are not claiming to be "uncovering reality"; rather, we maintain that this method, the scientific project of radical behaviorism, is a method that works for what we want to do. The pigeon in Skinner's experiment could say something similar: "Pecking this spot works when it comes to getting Skinner to give me food."

When we radically apply the fundamental principles of behavior that we have identified, this leads to another important result, one involving the definition of the term "behavior." In everyday speech, the word "behavior" normally refers only to external actions, which can be observed by anyone else who is present. So how should we regard the things a person does but that no one except the person himself can observe, things like feeling, remembering, and thinking? Traditionally, these phenomena have been assigned to another sphere—the psyche—as if they were of a different nature than the things we can observe. Here too, Skinner called for consistency, maintaining that there is nothing to indicate that the same principles are not valid for these phenomena as well (Skinner, 1953, 1974). This means these phenomena are also behavior, and that they can and should be analyzed according to the same principles as behavior that is observable by others.

FUNDAMENTAL BEHAVIOR ANALYTIC PRINCIPLES

What, then, are the fundamental principles demonstrated by and studied through experimental research that we can use to understand and influence behavior? For a more detailed answer to this question, the reader is referred to other publications (Catania, 2007; Ramnerö & Törneke, 2008). Still, I will provide a short summary here, before turning to this book's main quest, which is to point out how these principles should be used to shed light upon the function of human thinking.

The two fundamental principles for behavior analysis are operant and respondent conditioning. The latter has been described since Pavlov's well-known experiment with dogs at the beginning of the twentieth century, including how their natural reaction of salivation can be influenced through conditioning. Operant conditioning is the principle of learning that Skinner investigated and demonstrated via his experiments, so that is where I will begin.

Operant Conditioning: Learning Through Consequences

Human actions never take place in a vacuum. There is something preceding and something following each action. It is among these contextual factors—those that precede and those that follow—that the behavioral analyst looks for answers to questions about what governs behavior. If someone, in a certain context, turns his eyes to me with a specific expression on his face, I might address him by saying something like "Can I help you?" My utterance is followed by a new occurrence: The person replies. So my action is followed by a consequence, in this case that someone answers me. The core principle in *operant conditioning* is that the consequences following a behavior (a response) influence the probability of the behavior being repeated. Let's speculate a bit using two rather different consequences in the everyday example above. Imagine that what follows upon my utterance "Can I help you?" is that the person gives me a friendly smile and tells me what he wants. An alternative would be the response "Mind your own business, you jerk!" It is hard to say exactly how each of these consequences might influence the probability of me asking if I could help if someone else were to look at me with this specific facial expression in the future. What is essential is that earlier consequences do have an influence. "A burned child shuns the fire" is an old saying in Swedish, corresponding to "Once bitten, twice shy" in English. A well-known author has turned this "truth" around in a book entitled *Burned Child Seeks the Fire* (Edvardson, 1997). It is not always easy to determine what kind of behavior to expect based on previous consequences. But the saying and the author agree on one thing: Previous consequences have an influence. This is the core of operant learning.

In operant psychology, the different influences of consequences are categorized based on whether they increase or reduce the probability of an earlier behavior being repeated. If I get friendly responses to utterances like the one above and thereafter more often address people who exhibit the same type of facial expression in similar situations, the friendly responses would be said to have had a reinforcing effect on this specific behavior of mine. A consequence that increases the probability of the preceding behavior being repeated is thus termed *reinforcing*. In this case the reinforcing consequence is that I receive something: a friendly response. Something is added. This kind of process is called *positive reinforcement*.

A behavior can also be reinforced through a consequence consisting of something being removed. This is illustrated by the behavior of the person who responded to my question by saying, "Mind your own business, you jerk!" Let's assume that when this is uttered, I become quiet and turn my attention away from the speaker. I do what I am told to do. This consequence could increase the probability of the other person repeating the utterance "Mind your own business" in similar situations in the future. The consequence—that I became quiet and turned away—has in this case become reinforcing to this person's behavior of telling off jerks. This time, though, the reinforcement consists of something being removed, namely, the attention from a jerk. When a behavior increases because something is removed, it is termed *negative reinforcement*.

Distinguishing between positive and negative reinforcement (which are both processes that increase the probability of a certain behavior) is not always essential. These two concepts can be said to describe two different sides of the same thing (Michael, 1975). If the behavior of telling off jerks was reinforced by me becoming quiet, then one factor is that my annoying questions came to an end. This is negative reinforcement. But another way of describing the same thing is to note the condition that resulted: silence, for example, or anything else that was added. That would be positive reinforcement. It is often convenient to distinguish between positive and negative reinforcement, even though the difference may not be clear-cut from a theoretical perspective. It may sometimes be more obvious that something is removed than that something is added. By speaking about this as negative reinforcement, we clarify the process. The distinction is often practical in clinical situations, a subject I will return to in part 3 of the book.

When a consequence reduces the probability of a certain behavior being repeated, it is known as *punishment*. If I receive an unfriendly response in the example above and I subsequently refrain from addressing people in that specific social context or do so less often, then the earlier consequence has been punishing. Punishment, too, can be separated into *positive punishment*—in which something has been added—and *negative punishment*—in which something has been removed. Remember, however, that there is no way of determining what is reinforcing versus punishing based on any intrinsic quality that signifies the consequence as such. Of course, it is true that some consequences more often function as reinforcing to people, for example, certain types of social attention. But this is not always the case. To be recognized and addressed in a kind way is usually reinforcing for human behavior, but we can all think of situations when this is something we want to avoid. Likewise, certain consequences usually function in a punishing way, like being hit, yet this is not always the case. There are situations when being hit reinforces the behavior that preceded this consequence. A child who encounters only indifference, despite several actions meant to attract attention, may repeat a behavior that leads to getting smacked, simply because the smack involves attention. It is the *function* of the consequence that provides the definition. When the consequences increase the probability of a certain behavior, this is reinforcement; and when the consequences reduce the probability of a behavior, it is punishment. To recap:

DIFFERENT TYPES OF CONSEQUENCES

Reinforcement: A consequence that increases the likelihood that a certain behavior will be repeated.

- **Positive reinforcement** is when the consequence is something that is added.
- **Negative reinforcement** is when the consequence is something that is taken away.

Punishment: A consequence that decreases the likelihood that a certain behavior will be repeated.

- **Positive punishment** is when the consequence is something that is added.
- **Negative punishment** is when the consequence is something that is taken away.

Before I go on to describe how we can analyze a specific behavior, I must clarify what it is that we are analyzing. Two behaviors are rarely, if ever, identical; even if they seem to be, they differ in detail. I can raise my coffee cup in a number of different ways, and I can address another person in many different ways. In behavior analysis, behaviors that are alike in the sense that they have the same function are said to belong to the same *functional class*. This categorization is essential if we are to analyze behavior. When you analyze a given, historically defined behavior, it is of course this particular behavior that you are analyzing—for instance, the way I addressed someone in the above example. At the same time, this is only of interest if it is useful when analyzing a similar behavior in the future—a behavior that is similar enough to have the same or almost the same function. Over time, we are primarily interested in functional classes or categories of behavior. Some such classes of behavior are narrow or very specific, such as the behavior of a biathlon competitor in erect shooting position when he aims to score a bull's-eye. Other classes are broad or contain a wide range of behavior, such as what people do to avoid painful memories.

ABC

In behavioral analysis, a common way of describing an operant sequence of events is ABC, and analyzing such a sequence is referred to as carrying out a *functional analysis*. The core is B, which stands for "behavior": Something is being done. This behavior or response is what we intend to predict and influence. C stands for "consequence"; we have just seen the significant role it plays in an operant analysis. Finally, there is A, which stands for "antecedent," or "that which precedes." Even the consequences that govern a certain behavior actually precede the behavior they govern, because they have followed an earlier, similar behavior. If speaking to another person has previously been followed by receiving kind attention, the likelihood of this behavior being repeated may increase. However, the antecedents referred to as A in ABC are those conditions that are present when a behavior occurs. Within behavioral analysis, we think of A as having at least two different types of function: discriminative function and motivational function. I will begin by describing discriminative function.

The fact that my question ("Can I help you?") was reinforced earlier does not mean that from then on I will always be asking the same question. This behavior has been reinforced in a specific context, and it is in this context—or rather, in similar contexts—that the likelihood of my asking the question again increases. Antecedent refers to precisely this context: the conditions under which my behavior previously put me in contact with certain consequences that have become governing. It was under the condition of having a person look at me with a certain facial expression (A) that I spoke to the person (B) and as a result of this behavior encountered different consequences (C). This means that A now has a function established by a certain connection in my history—a connection between a condition, a behavior, and a consequence. Now, when I meet a new condition that is similar enough to the one I encountered earlier, the previous consequence influences my current behavior. This function of A is termed *discriminative,* and when this is the function we have in mind, we speak of the antecedent as a *discriminative stimulus.*

A discriminative stimulus signals a historical connection between a behavior and a certain consequence. A specific type of behavior in another person—a glance or a facial expression, for example—signals to an individual a historical connection between a certain behavior, like asking a question, and a certain consequence. You might say that a discriminative antecedent in the present signals the availability of a certain consequence, simply based on this historical contiguity. This connection is called a *contingency* in behavior analytic language. We also say that a certain consequence needs to be contingent on a certain behavior in order to exert its function. What we mean is that there must be a direct connection between the behavior and the consequence. In order for discriminative antecedents and reinforcing or punishing consequences to have their respective functions in relation to a specific behavior, they must occur in contiguity with that behavior. A great deal of research has been carried out in the field of experimental analysis of behavior to try to describe these connections more explicitly, as well as how they may vary (Catania, 2007).

The prevailing circumstances (A) can affect the probability of a certain behavior in another way too. There are circumstances that are not discriminative; that is, they don't signal any historical connection between a behavior and a certain consequence. Although they don't signal an increased availability of a specific consequence, they influence the likelihood of a certain behavior. A classic example is food deprivation (hunger). If my daughter walks past the kitchen, where I am cooking a meal, and I say, "Dinner is ready," this might function as a discriminative stimulus for her. If so, she stops walking toward the TV room and sits down at the kitchen table. There could be different types of learning history behind this, but one possibility is that my words signal to my daughter a historical connection between the current conditions and the availability of a certain consequence: that of being served food. This is a description of a discriminative function.[1] Alongside this, however, the fact that my daughter is either full (after just eating a couple of sandwiches) or hungry (because she has not eaten since breakfast) may influence whether she sits down at the table to eat. Her being full versus being hungry does not indicate the availability of food. The meal is equally available regardless of whether she is hungry or not. Her hunger is representative of a different function of conditions that may precede and influence a behavior. This function is usually called an *establishing operation* or *motivational operation* (Michael, 1993). It is a function of A, but not a discriminative function. These are antecedent conditions that influence the reinforcing or punishing effectiveness of a consequence. In this example, the dinner will be more or less reinforcing to my daughter depending on her recent history with food—that is, whether she is hungry or not.

Let's take a look at the earlier example of asking "Can I help you?" with this aspect in mind. We could assume that I have a history in which a certain behavior by another person (facial expression, for example) functions as a discriminative antecedent for asking a question. Let's imagine two possible scenarios: I could be unusually tired because I didn't sleep well the night before; or, on the other hand, I may have been missing social contact with others for some time. Both of these conditions could influence whether I ask my question or not, if I indeed encounter a discriminative stimulus for this behavior. And this is despite the fact that neither my tiredness nor my desire for human contact indicates any availability of the governing consequence. Instead, the issue here is that these conditions influence the degree to which the governing consequence is motivating to me, or how much influencing power the consequence has in this particular situation.

---

1 As I will go on to describe later, this is a simplification that does not take into account the difference between verbal and nonverbal discriminative functions. The example works for the intended point in this case, though. There are quite a few of these types of simplified examples in this chapter. The alternative would have been to only use examples from organisms without human language, but this would have affected the text in a negative way and would hardly have helped the reader's understanding. The problem with these types of simplifications, which have been necessary within the area of behavioral analysis due to its difficulty in handling phenomena like language and cognition, will be dealt with in detail later in this book. It is also important to remember that within behavior analysis we are aiming at usefulness, not necessarily at covering all possible aspects of an event (see Ramnerö & Törneke, 2008).

BASIC FUNCTIONS OF THE DIFFERENT PARTS IN AN ABC ANALYSIS

A	B	C
Antecedent	**Behavior**	**Consequence**
Discriminative function		Reinforcing function
	Something an individual does	
Motivational function		Punishing function

It is often useful to distinguish between the discriminative and motivational functions of the conditions that precede and influence operant behavior. But again, it is the degree of usefulness that determines how essential this distinction is. In practice, it is not always possible to make the distinction, and other times it may not be important to do so even if the possibility is there. This leads me to an important point about what ABC is not.

A NOTE ON WHAT ABC IS NOT

When we describe a behavioral sequence in the way I have above and assign different terms to the different functions, it is easy to be misled in thinking that we've discovered a mechanical chain of events "out there in reality." But, as mentioned earlier, this is not what is intended. This type of discussion is simply a way of speaking and writing about behavior—a way that is useful for gaining both understanding and influence. In the constant flow of events we face, we can distinguish between different processes because it is useful to us to do so. This applies to all humans at all times, even those engaged in scientific study. Distinguishing between different processes, principles, antecedents and consequences, reinforcement and punishment, and so on is a behavior performed by the behavior analyst. Within this frame of reference, knowledge is seen as a skill for doing certain things; it is not an object that you discover or find yourself in possession of. Analyzing behavior from the perspective of radical behaviorism is not uncovering a hidden reality. Everything we do is acting on reality, or behaving. This means that our activity of performing ABC analyses is operant behavior, as well, governed by the consequences that the behavior analyst has previously met when doing this very activity. We do what we do because we have a history of connections between different antecedents, earlier behaviors, and consequences. We do it because it helps us reach certain goals.

From this position, making claims regarding discovering or understanding "the nature of reality" is a self-contradiction. Being based in functional contextualism means we refrain from such claims of discovering or knowing the "truth." Instead, we adopt a *pragmatic truth criterion* wherein what is true is what is serviceable toward a certain aim or goal. This also means that a science needs to clarify its aims. Nothing works "in general." If it works, it works for something specific, for what we are aiming at. The behavior analyst has a twofold aim or purpose: prediction and influence.

Respondent Conditioning: Learning by Association

Whereas the power of consequences over behavior is the main point in what we call operant conditioning, *respondent conditioning* describes the power of certain antecedents to trigger a reflexive behavior. Put simply, under a certain circumstance we will react. If the same circumstance, or one much like it, reoccurs, it provokes the same reaction based entirely on the antecedent. The behavior occurs regardless of earlier consequences following this reaction.

Again, it is important to remember that this does not uncover mechanical processes that are actually out there "in reality," as phenomena in their own right. The terminology of behaviorism is simply a way of speaking about this issue, and we employ it because it is useful. It serves our purpose to distinguish between operant and respondent in this way. Although these processes coexist in the web of events we are trying to understand and influence (more on this below), for the sake of clarity I will isolate what we think of as respondent processes.

There are certain basic reactions that we do not need to learn. They are already there from the start. Loud noises, a physical blow, rapidly approaching movement toward our eyes, contact with strong heat, and so on—all of these trigger spontaneous movements in humans, as well as in other animals. Consequences of our behavior do not seem to affect this substantially. If we make sure that these spontaneous movements are followed by specific consequences that would typically affect operant behavior, this does not cause the corresponding change in behavior we might expect. If I encounter a certain consequence because of putting something in my mouth, this consequence will probably take on a governing function. Whether what I put in my mouth tastes good or bad will influence my tendency to put the same thing in my mouth in the future. This is operant behavior, governed by consequences. But if we were somehow able to make my salivation result in a bad taste, and then we did something that normally makes me salivate, this would not affect my salivation to any appreciable degree. Salivation is respondent behavior; it is a result of antecedents and is not significantly controlled by consequences. A response that has not been learned is called an *unconditioned response*, and a stimulus[2] (in this case an antecedent) that triggers such a response is called an *unconditioned stimulus*. Typically, salivation is an unconditioned response, and food is an unconditioned stimulus that triggers it. Some affective reactions, like fear, are also examples of unconditioned responses. These are all innate reactions that have evolved because of their survival value.

The fact that respondent learning occurs means that experience nevertheless influences both these reactions and their occurrences. This is true of both salivation and affective reactions, among others. Some external situations trigger fear without the individual having to learn this (Öhman, 2002). But in a situation where I feel afraid, other stimuli that are also present may take on or acquire the same triggering function as the stimulus that originally triggered fear. If I get assaulted while strolling in the main square of the town where I live, we can assume that I'd feel some kind of fear during the event. But if I later take another walk in the square, stimuli that had previously been neutral—or that had perhaps even triggered positive emotions in me—might instead provoke fear. The stimuli that have been conditioned in this way could be the square itself, the aroma outside the restaurant near where I was assaulted, or any of a number of other details, in themselves irrelevant, such as the statue standing in the square. As a result of these stimuli being present at the time of the assault, they are now associated with it. This type of learned response, called a *conditioned response*, is an example of how respondent reactions can propagate when neutral stimuli become associated with stimuli that

---

2 The term "stimulus" is common within behavioral psychology. It denotes a quality or a phenomenon in the context or environment surrounding a behavior (Catania, 2007). Thus, both antecedents and consequences can be called stimuli. A problem with this term is, perhaps due to the experimental history of the behavioral position, that it can easily be associated with something small and very clearly defined. This is not a requisite meaning. It can refer to something very complex and composite and be synonymous with what, in ordinary terms, we call an event. For this reason, in this book both words are used, with the latter ("event") often used to refer to the more composite meaning.

trigger unconditioned responses. In this way, formerly neutral stimuli become *conditioned stimuli.*

In respondent learning, the direct connection between events is important. A conditioned stimulus acquires its function by occurring in direct connection with an unconditioned stimulus and its attendant response. These connections and their variations make up another area that has been the focus of a lot of research (Catania, 2007).

RESPONDENT LEARNING

Unconditioned stimulus → **Unconditioned response**

Assault — *Fear*

↕

Neutral stimulus

The main square in my town

<u>After the above experience this might follow:</u>

Conditioned stimulus → **Conditioned response**

The main square in my town — *Fear*

Operant and Respondent Learning Interact

In many cases, it can be useful to distinguish between operant and respondent learning. Some processes are more easily understood or influenced based on operant principles, while for others respondent principles are more useful. However, learning often takes place under the influence of both principles simultaneously, and through interaction between them. If my son lives in a different city and I like talking to him on the phone, I may call him up once in a while. If it turns out that it's easier to get ahold of him on Tuesday nights, when I realize it is Tuesday night, I might call him. So far this is operant learning through positive reinforcement for calling, where Tuesday night becomes a discriminative stimulus, indicating the increased availability of a certain consequence: that my son will be there to answer the phone. Now, let's say that when I call my son on the phone, a specific melody is played while I wait for him to pick up. After I have encountered this on several occasions of calling, one day I hear the same melody on the radio. I start thinking about my son, and maybe some emotional reactions, originally occurring due to my interaction with him, also surface. How did this happen? The answer is respondent learning. The melody has become a conditioned stimulus, and thoughts and feelings connected with my son are a conditioned response. It is easy to see how these reactions can in turn function as antecedents for more operant behavior. I might, for example, call my son earlier than I would have if I hadn't heard this melody on the radio.

Another way in which respondent and operant learning interact is in how reinforcers and punishers are established. Stimuli that have not previously functioned as reinforcers can acquire this function by being associated with things that already function as reinforcers. Status symbols, designer clothes, a photo of a loved one, or a favorite TV program can all have reinforcing functions. These stimuli have acquired their function by being associated with other reinforcers; for example, receiving interpersonal attention because of status symbols. Reinforcers that have their function without being learned—such as interpersonal attention, food when you are hungry, and warmth when you are cold—are called *unconditioned* or *primary reinforcers.* Reinforcers that have acquired their function by way of learning are called *conditioned* or *secondary reinforcers.* The corresponding terminology also applies to punishers: They can be *unconditioned* or *primary punishers,* or they can acquire their functions through association with other punishers and therefore be called *conditioned* or *secondary punishers.*

Extinction

Behavior that has been learned does not necessarily last forever. Whether governed by consequences or by associations, behavior can cease or decrease following the removal of particular contingencies. We often use the term *extinction* for this; *operant extinction* or *respondent extinction,* respectively. Operant extinction occurs when a certain behavior no longer provides that which has been a reinforcing consequence. If my Tuesday calls to my son begin to go unanswered, I will probably continue to call him on Tuesdays for a while. But if I encounter that he never answers on Tuesdays anymore, I'll stop calling him on that day. Tuesday is no longer a discriminative stimulus for the consequence that was previously reinforcing. Tuesday night no longer signals a historical connection between a specific behavior (calling my son) and a certain consequence (he answers the phone).

Respondent extinction also involves crucial changes or a discontinuation of the relations between the different stimuli that lay behind the learning. If, despite the respondently aroused fear I experience after being assaulted in my town square, I continue to regularly visit the square, my fear will decrease, provided that the event (the assault) that originally caused my reaction is not repeated. New associations between my emotional reactions and the square, the aroma from the restaurant, and the statue are established. I may experience other events in the square that awaken positive emotions, so new respondent learning takes place. That this happens when I walk in the square (operant), even though I feel afraid initially (respondent), is yet another example of how operant and respondent learning act together.

That a particular behavior is extinguished does not mean it has been unlearned. Once I have become afraid in connection with the assault in the square in my hometown, this fear will most likely be more easily triggered than it would have been if I had never experienced the assault in the first place. If my son's phone does not play the special melody any longer, then the melody itself may not bring up thoughts of him if I hear it on the radio. And yet we've all experienced how associations like these can suddenly reemerge after being gone for years. The same is true for operant extinction. Actions we once learned often lie within reach, even if we have long since ceased to act in that way. Although I once learned to ride a bike, this behavior might have been extinguished many years back if the reinforcing consequences ceased. If I get on a bike again, I might notice that I cannot ride it as well as before; nevertheless, this behavior is not unlearned.

The fact that learning has these kinds of lasting effects tells us something important about working for change. I will return to this point when discussing this book's main issue: the power of thinking. What we can establish at this juncture is that the key is to learn new things, rather than trying to extinguish what we have already learned.

Generalization

The fact that, in an operant sequence of events, a certain stimulus functions as a discriminative or motivational antecedent, or as a reinforcing or punishing consequence, does not mean that a new event must be identical to have the same function. If that were the case, learning would practically be impossible, since two events are, in fact, never exactly the same. Instead, two stimuli or two events need only be "similar enough." If a certain social behavior from my side, like addressing someone with a question, has taken place under the condition of this person looking at me with a certain expression on his face, and if my asking the question has led to positive reinforcing consequences, then this increases the likelihood that I will speak to someone again when the conditions are similar; the situation need not be identical. How similar the conditions must be in order to function in a discriminative way depends on the individual's specific learning history. A young child initially may have learned to say something in an interaction with a parent. This is the condition under which the behavior has been reinforced. A glance, a facial expression, or an utterance by a different person does not at this stage constitute an antecedent for the child to interact in a similar way. This antecedent function will, however, gradually spread, becoming *generalized*. As time passes, a much broader category of conditions (different people, different environments, different utterances, and slightly different facial expressions) might function as antecedents for a certain social behavior.

This does not apply only to antecedents. Events that function as reinforcing or punishing consequences are generalized, as well. Friendly behavior may be encountered in many different forms, but regardless of the differences it can still have the same reinforcing function. The same goes for the things that often function as punishers, like being criticized, for example. Another example of a reinforcing function that can be generalized involves money. Money is a conditioned or secondary reinforcer. In our experience, it has been associated with other things that have been reinforcing, giving these pieces of metal and paper reinforcing functions in themselves. Money has acquired this function by association with such a large number of other reinforcing functions that it becomes a generalized reinforcer. For most children, attention from adults also functions as a generalized reinforcer. In certain contexts, however, the same attention can have a punishing function. This shows us, again, that whether a given stimulus is reinforcing or punishing is not an intrinsic quality of that stimulus; rather, it can only be understood in context, in the interaction between an organism and its environment.

Generalization is relevant in connection with respondent learning, as well. The fact that I was assaulted in the main square of my hometown can make me feel this fear in a similar square located in a different town if that other square is similar enough. Or this fear can emerge in a totally different situation from being in a square; for example, seeing someone in the subway who is somehow similar enough to the person who assaulted me. Likewise, hearing that familiar melody on the radio can make me think of my son even if it is performed by a different singer and in a different version than the one on his phone.

Discrimination

Discrimination can be said to be the opposite of generalization. Just as the function of an event can be spread to other events because they are in some way similar, a function can be restricted to a more specific event in situations where a similar event doesn't have this function. Another person's facial expression can be an antecedent for a certain social behavior on my part, and we might assume that generalization has taken place in my learning history since my first attempts at interacting with other people years ago. There are quite a few slightly different facial expressions of other people that lead to more or less the same type of social behavior from me. But what happens if I want to play poker? In that case, the very small variations that I might be able to detect in the other players' expressions become antecedents for quite different behaviors from my side. A certain glance may make me raise my stakes, and if another player looks at me in a slightly different way, this can be an antecedent for my decision to fold. The differences could even be so small that I'm unable to describe them, yet I still act on them. The fact that some people are so much better than others at playing poker can partly be assumed to be connected to a highly trained ability to discriminate when it comes to other people's behavior.

Generalization and discrimination, and the balance between them, is an important part of all kinds of learning, both operant and respondent. Sometimes we need to catch one very specific signal in a noisy surrounding, while at other times it is important to act on anything that moves.

WHERE WE GO FROM HERE

This concludes my run-through of important foundations of the psychology of learning. Let me once again encourage interested readers to seek more detailed knowledge from other sources. In chapter 2, we will move on to the issue of how these principles have been used and are used in order to understand thinking and the power of thinking. We will also take a look at some difficulties in relation to this—what can be described as inadequacies in the behavior analytic attempts to deal with this profound type of human behavior.

CHAPTER 2

Thinking and Human Language

The phenomenon we commonly call "thinking" is obscure to most of us. There is, of course, always an observer, but only one: the person thinking her thoughts. Thinking, then, is a phenomenon that is private and cannot be observed directly by anyone else. Thoughts are included in the phenomena Skinner termed *private events* (Skinner, 1953). However, we might critically remark that thoughts are nevertheless often accessible, like when something is said and then heard, or is written and then read. If we think about it carefully, though, we find that what is accessible isn't the thought itself; it is what is said or written. These expressions are often related to the original thought, but they are hardly identical to it. The experience of thoughts in themselves is directly available only to the person who is thinking those thoughts. The same truth applies to recollections, internal bodily sensations, and at least parts of what we call emotions, all encompassed in Skinner's term "private events." It is important to emphasize that to Skinner, this did not imply that "internal" phenomena are of a different nature than what we can plainly observe jointly, or publicly. These phenomena are not enacted in a mental world that is somehow different from the external world. The distinction is purely in their degree of accessibility to public observation.

We humans devote a lot of attention to these private events. We talk about them and assume that they are present in others, even though we are unable to observe them in anyone but ourselves. We attach great importance to them in a number of different ways. What is just now taking place is that I, as an author, am writing a book about some parts of this phenomenon. You, as a reader, are devoting your time and attention to understanding what I am writing, and you may be thinking (if I may dare guess) something along the lines of "It will be interesting to find out what his thoughts (!) are on this." This phenomenon of thinking is also given a prominent position within modern science's understanding of human beings, usually under the heading of "cognition."

In his interest in private events, Skinner again showed his consistent, or radical, stance, questioning why we talk about these private events. This is definitely a very essential question to ask. If we understand why we talk about this, we will know something about the function these phenomena have for us. In addition, if we did not learn to talk about things that only the individual can note, these phenomena would hardly turn into what they gradually become to an adult person. The fact that we do things jointly in connection with these internal phenomena gives them a function for us. In fact, these phenomena attain a center-stage place in human behavior precisely through known principles of learning. We learn to talk about private events, and the way we learn to talk about them, in turn, influences how they evolve. (What I discuss as "talk" is not limited to producing sounds; for example, a mute person can learn the same thing through sign language.)

Why would it be important to humans to learn to talk about feelings, memories, physical sensations, and thoughts? Because this kind of sharing is valuable to the social community. In Skinner's words, "It is only when a person's private world becomes important to others that it is made important to him" (Skinner, 1974, p. 35). Expressed in a slightly more technical way, we learn to talk about the things each of us can observe in only ourselves because our social environment reinforces that type of behavior. The social community reinforces children for speaking based on their own private events. Step-by-step, these phenomena—which the social community cannot observe in a direct way—become important to the child, as well, based on the contingencies of reinforcement.

VERBAL BEHAVIOR

Before we take a closer look at how speaking about private events evolves, we need to reflect on speaking in a more general sense: what Skinner called *verbal behavior*. From very early in language training, humans learn to use combinations of sounds in a way that successively becomes very important for how we interact. What is functionally crucial in this behavior, according to Skinner, is the possibility of behavior being reinforced in an indirect way, by how other individuals act, rather than as a direct result of the speaker's actions. This makes it possible for me to be given a desired object without having to direct my actions toward physically obtaining the object myself. I can ask for something and receive it by way of another person's actions. Using sounds in this way is primarily a social ability, and it is carried out according to the same fundamental principles for learning that apply to other behavior. Skinner called this kind of action by a speaker "verbal behavior" and defined it as a behavior that acquires its effect through the mediation of a listener's behavior, the listener having been taught such that her behavior precisely functions as reinforcing for the behavior of the speaker (Skinner, 1957). If I reach for an object, the consequence of managing to reach it is reinforcing for that behavior. Asking to be given the same object is only reinforcing if it is followed by the listener's behavior in the form of giving me the object I wanted.

Someone speaks, and based on this speaking, another individual acts in a way that makes different reinforcing consequences accessible to the person who spoke. This also allows the social context to govern the speaker's behavior in turn, based specifically on the contingencies of reinforcement that are established. If the listener, when she hears me, gives me what I asked for, the probability of my repeating this behavior increases. If the listener acts in a different way, this too will influence my future behavior.

SKINNER'S DESCRIPTION OF VERBAL OPERANTS

Skinner divided verbal behavior into several primary types: tacts, mands, echoic behavior, intraverbal behavior, and autoclitic behavior (Skinner, 1957). When we regard verbal behavior as operant responses, we see it as controlled by antecedents and consequences, just like any other operant behavior. These different responses (B in an ABC analysis) are distinguished by the different relations between the form (what is said or written) and the variable (A and/or C) that governs the response. The shape the response takes and its relation to antecedents and consequences is what forms the basis for classification.

Tact

A *tact* is governed by a preceding stimulus: the stimulus that is being tacted. An example would be saying "chair" when a chair is present. This response, the utterance, is a direct result of seeing the chair. When we say "She is running," it is governed by the fact that someone (she) is moving her body in a certain way. We tact our environment in this way because we have a solid learning history in which tacting has been reinforced. From an early age, we have experienced reinforcing consequences when, for example, in the presence of a cow, we have uttered precisely "cow." If we said "kitty" in the presence of a cow, there were other consequences. The governing consequences are primarily of a general and social nature; that is, tacting is followed by generalized reinforcers. It is tacting we have in mind when in everyday speech we speak of "describing," "telling," "referring to," and the like. All of these concepts, however, are highly imperfect for scientific purposes, and thus the use of the neologism "tact."

An ideal tact is completely controlled by the stimulus preceding it. In an everyday context, we would say the statement is objective or corresponds to the object referred to. This is the type of "pure" tact that we seek in scientific linguistic inquiry. Tacts rarely have this character in normal life, and you could probably question whether a pure tact is actually possible as anything other than an ideal, even in scientific settings. Skinner wrote about distorted or impure tacts (1957), by which he meant tacts that are controlled by other factors, such as who is listening and how listeners act as a result of a given tact. What we normally speak of as exaggeration is an example of a distorted tact. The verbal operant is controlled by that which precedes it, but the size, for example, is exaggerated, so the tact is distorted. When the "Great Fisherman" talks about the size of the fish he has caught, his descriptions are governed not solely by the size of the fish. If they were, the descriptions would be pure tacts. Instead, they are probably governed by other elements, as well. If a statement is portrayed as a tact when it actually is governed not at all by what precedes it, but rather by something completely different, in everyday language we would call it a lie.

Mand

A *mand* is verbal behavior controlled by a specific reinforcer, and it specifies this same reinforcer. For example, your saying to a person "Go away!" is reinforced by earlier experiences of having a person leave after you said this. Saying "Look at me" specifies its own reinforcer—that is, the listener looks at the speaker. The typical antecedent (A) of a mand is the presence of a listener (a discriminative function) and a motivational operation (an establishing operation) that makes the consequence in question desirable to the speaker. The motivational operation that precedes "Go away!" is most likely that the listener's presence is aversive to the speaker at the moment, whereas with "Look at me," something else is going on in the interaction between speaker and listener. Examples of mands are different types of requests and demands, asking questions, and raising your hand to obtain permission to speak.

It is important to understand that both tacts and mands are defined by their function, not by sheer topography. The very same word or expression can have several different functions. This means that topographically identical linguistic expressions can function both as tacts and as mands, depending on the relation to antecedents and consequences in the specific situation. When someone says "the newspaper," it could be a tact if the governing variable is an actual newspaper and the utterance is an answer to the question "What is that on the table?" But "the newspaper" could also be a mand if it serves as a request that someone hand the newspaper to the speaker. The relation between what is said and the governing variables determines what type of verbal behavior is at hand. A statement such as "Those apples are nice" can seem like a tact but in fact be a mand, if this statement is reinforced by the fact that the consequence that previously followed upon a similar statement was that the speaker was given an apple.

Echoic Behavior

In *echoic behavior*, verbal behavior has an antecedent that is topographically identical to the response. It is a verbal response that follows a preceding verbal response, echoing or repeating something that has been uttered. Again, typical reinforcers are generalized social consequences like attention and other interpersonal processes. Echoic behavior is a core part of early language learning. The parent, for example, utters a word and then reinforces every tendency in the child to repeat it. But echoic behavior continues to exist as an important verbal behavior throughout life, like when we silently repeat something we have just heard.

Skinner gave a few more examples of verbal responses that resemble echoic behavior insofar as they all involve a response that is somehow a reiteration of the antecedent. He described textual behavior, as well as transcription and taking dictation. I describe all of these together with echoic behavior based on this similarity. *Textual behavior* is saying something that is controlled by a preceding stimulus in the form of a written text, where a formal correspondence exists between what is written and what is said. Textual behavior, then, is the response we would normally refer to as reading aloud. *Taking dictation* is an inverse process: writing down something that is in formal correspondence with what has been uttered, like when you write down a telephone number someone just told you. *Transcription* is writing something wherein the controlling antecedent is topographically identical with the response. In everyday language, this is called copying. All of these verbal behaviors are reinforced by generalized social reinforcers.

Intraverbal Behavior

Intraverbal behavior is also verbal behavior that has other verbal behavior as its antecedent, just as echoic behavior; but in this case there isn't any formal correspondence between antecedent and response. In this case, the relation between the verbal antecedent and intraverbal behavior is arbitrary, established by social whim. If I say "one, two, three," and you say "four," then your response is intraverbal behavior. If I say "What is *casa* in English?" and you say "house," then this response, too, is intraverbal. Just as for all other verbal operants except the mand, the important, governing consequences for intraverbal responses are generalized social reinforcers.

Once again, note that the definitions of these different types of verbal behavior are functional. What is crucial is the relation between the response and the antecedents and consequences. Above, I said that the expression "the newspaper" can be either a tact or a mand. But it could, of course, also be echoic behavior if it were governed by someone else saying "the newspaper" and the response is a reiteration of this. It could also be intraverbal behavior. This would be the case if it were governed by someone else just having said, "What is another word for 'the local rag'?"

Autoclitic Behavior

Skinner described one more type of verbal behavior: *autoclitic behavior.* This is verbal behavior, or parts of verbal behavior, governed by other verbal behavior by the speaker and modifying this other behavior. An example of autoclitic behavior is the word "maybe" in the response "Maybe it's the newspaper" when someone asks, "What is that on the table?" The word "maybe" functions as an autoclitic because it modifies the totality and lets the listener know something about the position from which the speaker is speaking, in this case a position of uncertainty. Other autoclitic behavior, which would modify the tact in a different way, would be a word like "not," as in answering the question about what is lying on the table with "not the newspaper." Autoclitic behavior can be whole words, as in the examples above, but it can also be modification of a word, such as adding an "s" to the end when the answer is "newspapers." Punctuation, grammatical structure, and syntax are all forms of verbal behavior in the category autoclitic behavior—verbal behavior that is dependent on or modifies other verbal behavior by the speaker.

The Use of Skinner's Analysis in This Book

Skinner's system for analysis of verbal behavior is fairly complex, and my intention has not been to describe it in detail here. With respect to the preceding outline of the analysis of verbal behavior, two points are important from the perspective of this book. The first is that this outline shows us that a functional analysis of verbal behavior is a feasible task. Verbal behavior is a human behavior that is governed by antecedents and consequences. The second point is that this analysis allows me to return to a consideration of private events. In this, I will primarily use one part of Skinner's analysis—his description of the behavior of tacting—to aid in understanding how private events attain such a central function for us humans.

LEARNING TO TACT PRIVATE EVENTS

What does all of this mean in terms of our initial question about how we learn to talk about internal phenomena? One thing it shows us is that we learn this from our social environment, which reinforces certain behavior. For the child, this environment is initially made up of parents, other guardians, or other family members. Hence, from a Skinnerian perspective, these members of our social environment are the ones who train us to appropriately tact events in our environment. However, these people (and others) do not have direct access to a child's private events. When people in the child's social environment reinforce tacting and what is to be tacted is an external phenomenon, like an object, a person, or an external action, the social environment does have access to whatever is to be tacted. The connection between stimulus and tact is obvious. Establishing a connection to a reinforcer is rather easy in this case. The child who is learning to tact a teddy bear can see the teddy bear in front of her eyes, and the same goes for the parent. When the child says "teddy" or something similar, the parent can reinforce the behavior. What the child observes internally, however, is not accessible to the social environment in the same way. This will make it more difficult for the environment to reinforce her tacting of such internal phenomena. Skinner described several ways in which people in the social setting can reinforce tacting of private events, despite this difficulty (Skinner, 1945).

One way we learn to tact private events is when others are able to observe phenomena that are parallel to what is being tacted. If others are able to observe a flush or a swelling in a child's skin, they can assume that the child experiences pain. In such a situation, the parent can reinforce verbal behavior that tacts the assumed pain, such as if the child says "ouch" or "hurts." The child's verbal behavior is followed by reinforcing consequences, established by the surrounding social environment. Other common, and commonly accessible, phenomena are events that we know normally produce feelings. One example would be when someone in the child's environment behaves in an aggressive manner, and we afterward ask the child, "Did that make you feel scared?" Having learned to tact our own private world, we use our own experiences and assume that the child experiences something similar to what we experience. However, a parent's emotions may not always correspond with those of her child, which leaves open the possibility of problematic tacting of emotions on the part of the child.

A similar type of learning occurs when observable responses by the individual are parallel to the private phenomenon that is tacted. Sounds, facial expressions, and certain movements are observable responses that commonly parallel an individual's private events. Experiences of pain, anger, interest, and joy are often accompanied by other behavior. Children shy away, draw closer, cast glances, and act in a number of different ways that are visible to others in their social environment. The fact that these different forms of observable behavior vary in accordance with private events makes it possible for the environment to establish contingencies of reinforcement that, from then on, influence the child's growing ability to talk about the things that only she can observe.

Another way of learning to talk about private events occurs when the child first learns to talk about the things she does that are observable both to herself and to others around her. She then gradually goes on to learn to talk about similar actions of her own that are observable only to herself. This is especially relevant to what we commonly call thoughts or thinking, and the same learning path is also relevant to what we normally call memories.

A child carries out a number of actions that she gradually learns to talk about. She walks, waves, eats, watches the dog, stands still, plays, gets dressed, and so on. No child can tact her own action without first having performed it. (She can, however, utter a word that others use to describe something that is performed without performing it herself. In this case, she's echoing something she heard someone else utter.) In order to tact her own actions, the following sequence is required: The child does something that is observable to the child herself and to her social environment. When this action is carried out, or immediately thereafter, the environment reinforces a certain verbal behavior by the child. Most likely, the child first repeats after others (echoes), then gradually transitions to tacting. The child's own action then becomes a discriminative stimulus for the verbal behavior.

Here's an example: Petra is kicking a ball. When she does this, her father says, "Kick! Look, you're kicking the ball!" The word "kick" is repeated, and if Petra says something that resembles "kick," her father reinforces this utterance. In time, the very action of kicking becomes a discriminative stimulus for Petra to say "kick." Petra has learned to tact her own kicking. Over time, similar learning occurs for other actions: waving, running, playing, and so on.

Alongside this process, the child will, of course, experience aspects of these behaviors that are private, that is, only accessible to the child herself; for example, what her arm feels like when she waves to someone. This will contribute to an increase in her ability to tact private events. She learns to tact her own actions, and because these actions vary in accordance with private events, her ability to discriminate private events gradually grows, along with her ability to tact them.

Skinner described yet another way in which we learn to tact private events, but before we go on to that, let's first take a closer look at tacting of the type of private events that are especially important as far as this book is concerned. This has to do with how we learn to think, how we learn to talk about our thinking, and how, in time, we learn to think about our thinking.

Learning to Tact Thoughts

An action that is initially accessible both to the person who performs it and to the social environment can gradually change (Skinner used the term "weaken") so that it is no longer accessible to others, although it stays accessible to the person performing it. One simple example is how we go about learning to count. First, children count aloud. That is the only way this behavior can be reinforced by the environment. As time passes, the child begins to experience that counting aloud isn't reinforced in all situations; in fact, sometimes it is punished. People may find it annoying to have a child counting out loud, so the reinforcement ceases. The child then starts to count more quietly and realizes that she can count quietly enough that no one else can hear it. She may still be moving her lips but not be making sounds. A child who is sitting quietly with a counting book and moving her lips in a concentrated way may also receive reinforcing social consequences for this behavior, and—presto!—she has learned to count silently, to "count in her mind." Provided that this behavior results in consequences that are functionally reinforcing, the behavior will continue. Since the ability to count silently is very useful to attain a number of things that people want, this remains a human ability. The same sequence is easy to see in the process that leads to our ability to read. First we do it aloud, and after a while we begin to do it silently. Still, a person who has been able to read silently for many years will often read aloud again if what she is reading is particularly difficult, like a complicated, unfamiliar word, even if she is speaking only to herself. And when we want our thoughts to be particularly effective, like when performing a demanding task, it is common to "think aloud": "Go for it, Niklas. You can do it!"

The preceding examples show how verbal behavior, such as counting or reading, starts out as a public event, where it is accessible to a social environment that can reinforce it, and then gradually turns private. A very large part of what we normally call "thinking" develops similarly. First we learn to talk about our own behavior based on the contingencies of reinforcement that have been established through the social setting; then, gradually, the public parts of the behavior wane while the private aspects remain.

I should perhaps point out that this is not a complete description of everything included in the terms "thoughts" and "thinking." It would be impossible to give such a description, for many reasons, including the ambiguity surrounding what should actually be included in the terms. They can be defined in different ways, and they are everyday concepts, not scientific terms. What I am maintaining, however, is that the process described above is fundamental to what we usually call thinking and thoughts. Also, this description is relevant to the aims of behavior analysis, that is, to predicting and influencing what we as human beings do, or how we behave.

Several factors contribute to the persistence of private verbal behavior, and to it becoming increasingly independent of those external actions in which it originated. Thinking about running is, in some ways, similar to actually running. It involves a number of internal stimuli identical to those that exist when you are in fact running. This is particularly evident when it comes to the kind of thinking that we call imagining, visualizing, or thinking of. When I think of my mother, I see something in my mind's eye that is very much like what I see when she is standing in front of me. The reality of such a connection between thinking and external behavior corresponds well with what we now know about the functions in the human body, and particularly the nervous system. Thinking about looking at someone is done largely with the same parts of the brain as actually looking at someone. If you first learn to play a piece of music and then hear this music played, motor neurons are active while you simply listen (Lahav, Saltzman, & Schlaug, 2007). Thinking about running involves the same motor centers of the brain as running (Jeannerod, 1994; Kosslyn, Ganis, & Thompson, 2001). This means that thinking about doing something is an action that in many respects resembles actually doing what we are thinking about. At the same time, this private action has great advantages. We can perform an action in the concealed setting of our imagination without facing many of the consequences that the external action would involve. Doing things privately—thinking of doing them—can thus be a way of testing and practicing. As we all know, this type of behavior plays a significant role in what we often call problem solving. We try things in our thoughts, and then we perform them more entirely, or else we refrain from doing them. It is easy to see that this possibility is likely to have increased our species' ability to survive.

The path of learning I have just described is thus:

1. We do something.
2. We learn to talk about what we are doing, which in the above terminology means we are tacting our own behavior.
3. We learn to speak without uttering any words; that is, we think.

Once we are doing this, this very behavior becomes something we can tact. Perhaps you, as a reader, just noticed that you were thinking about something other than what this text says. In that case, you have something new to tact: "I was just thinking..." Therefore, when we talk about the fact that we are thinking, this can lead to thinking about the fact that we are thinking. (In cognitive theory, this is often referred to as metacognition.)

Thus far, I have described two main ways in which we learn to talk about private events. The first way is when people in the environment observe phenomena that are parallel with private events in the individual, and they use these parallel phenomena to reinforce the verbal behavior for which the private events gradually become discriminative stimuli. The second way is when behavior that is initially accessible to the social environment gradually becomes private, through this behavior as a whole being punished or extinguished in certain contexts, while at the same time a part of the behavior—doing the same thing, only silently—leads to reinforcing consequences. The behavior as such can thereafter be tacted by the individual, just as other private events are. Before we turn to the question of why the social environment places such importance upon teaching each new individual this type of skill, let's look at a third possible way of learning the ability to tact private events.

Learning to Tact Private Events Through Generalization

Once tacting of private events becomes an established ability, generalization allows us to further develop this kind of verbal behavior. A private stimulus can have certain features that resemble other phenomena, either external or internal. Tacting something that is available only to the individual in question by using a resemblance to something that is accessible to others is a behavior that is often reinforced. One example is a child who has been sitting on the toilet for a long time and then gets up and says, "My legs are prickly like soda pop." The child has discriminated some sort of resemblance between the experience of drinking a carbonated drink and what she experiences in her legs in connection with having spent a long time sitting on the toilet. She uses what we could call a metaphor or an analogy to tact the similarity she experiences. The fact that a great deal of the language we use to describe private events consists of metaphors taken from events in our external environment illustrates this path for learning. When we use metaphors, we are making use of the similarity between different phenomena; thus, metaphors can contribute to generalization. Language like "simmering with anger," "feeling low," "being filled with peace," or "being overwhelmed" are all typical examples.

WHY IS THIS BEHAVIOR REINFORCED?

In the introduction to this chapter, I quoted Skinner, who wrote, "It is only when a person's private world becomes important to others that it is made important to him" (Skinner, 1974, p. 35). Why, then, is a person's private world important to the social environment, and what is the point in teaching a new individual to talk about this private world? The answer is that an individual's private world contains things that are useful in social interaction. We are social animals, and social interaction is fundamental to our survival.

What an individual feels in a certain situation communicates a lot about her history in connection with similar situations. For example, if someone says she is hungry, this is a brief way of describing her history in regard to the availability of food. It also lets us know something about her state at present and what is essential to her in the current situation, and it let us know something about her inclination to act in the near future. It communicates that soon she will probably eat something or act in some other way in relation to food. The same goes for verbal behavior that describes anger, weariness, joy, and so on. If someone says she is depressed, others receive a condensed description that may be highly relevant to their own actions and to their ability to anticipate that person's further actions. The point is not that these internal phenomena are some kind of autonomous force; rather, the significance of such expressions lies in how the private events they describe are related to the individual's learning history.

Once the ability to tact private behavior is established, this ability also becomes valuable to the individual. To quote Skinner once more, "A person who has been 'made aware of himself' by the questions he has been asked is in a better position to predict and control his own behavior" (Skinner, 1974, p. 35). Being able to foresee and control one's own behavior naturally implies an increased ability to achieve things that are desirable to oneself.

Before I conclude this passage about the relationship between thinking and verbal behavior, I want to underscore, once again, that this is not a description of every aspect that could be included in the concepts of "thinking" and "thoughts." It could be argued that even before this learned ability of silent verbal behavior is established, as described above, there is some sort of rudimentary behavior in the child that might be called "thinking" (Vygotsky, 1986). It might also be argued that something like this is present in other animals besides humans, in one way or another. How we view this argument depends on what we include in the concept of "thinking." In any case, we know very little about this possible rudimentary capacity and what function it has. And regardless of this, something novel and revolutionary happens as the child's verbal behavior shifts from being solely public to also becoming private.

THE OBSCURITY OF THE INTERNAL WORLD STILL REMAINS

Despite the fact that we learn to talk about private events, it remains more difficult for us to talk about these events than about phenomena that are observable by others. All through life, it will be easier to describe to others how we go about driving a car or painting a picture than it will be to describe how we do our thinking or how we feel when we are sad. All of this is logical as seen from the preceding analysis. In situations where we have learned to talk about our pain, anger, or fascination, at least parts of our behavior haven't been observable to anyone but ourselves. This means we've had relatively limited support from the environment in learning to talk about these things compared to phenomena that others can more easily observe. Therefore, our ability to discriminate and speak about these private phenomena will forever be less developed and less precise than it is for other, more observable areas in our field of experience.

A SPEAKER AND A LISTENER

An important aspect of verbal behavior is that when it occurs, there is always a speaker and a listener. This is particularly obvious when the verbal behavior occurs with another person present, but there is a listener even when someone is thinking; it's just that in this case the speaker and the listener are the same person. The latter situation is of special interest in this book, but we will commence with what comes first in every person's history: the speaker as one person, and the listener as another. Skinner restricted the concept "verbal behavior" to the speaker. In his definition, it is the speaker who behaves verbally. The listener reinforces this behavior, but according to Skinner, this action should be understood as cohering with the same principles that apply to other forms of reinforcement. Skinner saw no reason to deem the listener's behavior verbal (Skinner, 1957). The listener can, of course, also speak in her turn, and that would be a verbal act, but the listening itself is something Skinner mostly left out of his analysis of verbal behavior (S. C. Hayes, 1991; Schlinger, 2008). Still, it is quite apparent that, in many cases, what Skinner called verbal behavior has complex effects on the listener. When someone says, "Please wait outside, and I'll be right with you," this is a typical example of what Skinner called verbal behavior. The point, of course, is to influence the listener, making her go outside to wait. If private events in the form of thoughts are powerful, they must be exercised in relation to a listener, even though in this case the listener and the speaker are one and the same. All of this gives us good reason to investigate Skinner's thoughts on the listener's behavior. He discussed listening as a separate category of behavior and referred to it as rule-governed behavior.

RULE-GOVERNED BEHAVIOR

"Please wait outside, and I'll be right with you" is a verbal statement that is easy to analyze using operant psychology's basic formula, ABC. If, as a result of this statement, the listener goes outside to wait for the person who made the statement to join her, we could analyze it as follows: The statement functions as an antecedent (A) for the behavior of going outside (B) in order for the speaker to join the listener outside (C). According to Skinner, the antecedent is functioning as a rule because it specifies a behavior and a consequence (Skinner, 1966, 1974). He emphasized the difference between behavior that is governed by such rules and behavior that is governed by direct experiences of consequences that have occurred, which he called contingency-shaped behavior. Here is a classical example of these two types of behavior, set forth by Skinner:

> The difference between rule-following and contingency-shaped behavior is obvious when instances are pretty clearly one or the other. The behavior of a baseball outfielder catching a fly ball bears certain resemblances to the behavior of the commander of a ship taking part in the recovery of a reentering satellite. Both move about on a surface in a direction and with a speed designed to bring them, if possible, under a falling object at the moment it reaches the surface. Both respond to recent stimulation from the position, direction, and speed of the object, and they both take into account effects of gravity and friction. The behavior of the baseball player, however, has been almost entirely shaped by contingencies of reinforcement, whereas the commander is simply obeying rules derived from the available information and from analogous situations. (Skinner, 1966, pp. 241-242)

Skinner went on to note that the sea captain can, by collecting many satellites, possibly step away from some of the rules, so that, in time, his behavior may gradually also become more shaped by contingencies. But rule-governed behavior has advantages, including that the captain does not necessarily need personal experience in order to succeed in his task.

It is easy to see that a great deal of human behavior is, in fact, rule-governed. "Don't drive through a red light." "Don't kick a man who's down." "Work hard on your studies, and you'll have a better chance of getting a good job." Some of the rules specify behavior and consequences that the individual probably has direct experiences of, as well, like "Put more clothes on, or you'll be cold." Other rules specify both behavior and consequences that the individual hasn't experienced, like "Stop abusing alcohol, or you'll die."

When we're learning to follow rules, they are given to us by people in our environment. But gradually we also learn to set up rules for ourselves wherein we become both the speaker and the listener: "Try again, and you'll do better!" "Leave it for now, and try again later." "Don't say a thing; that will just make it worse." Most of us have heard others say things like these to us, and we have also directed these sorts of statements to ourselves so that we might follow our own rules. And in precisely this formulating and following of self-directed rules lies much of what we commonly refer to as the power of thinking, which is the main theme of this book.

A DIFFICULT QUESTION

To utter a rule or an instruction is, by Skinner's definition, verbal behavior. However, as mentioned above, rule-governed behavior—following a rule—is not necessarily verbal behavior. If someone says, "Wait outside, and I'll be right with you," and I reply in turn, that would be verbal behavior. But if I simply go outside to wait, then, according to Skinner, it is not. Rather, it would be a result of previously experienced consequences and should be understood in the same way we understand contingency-shaped behavior. But this raises a difficult question: How is it, then, that we humans act with a view toward the future, toward things we haven't previously experienced, and that we do this as a result of something that has been said or thought? How can we understand the effect verbal behavior has on listeners in everyday situations like the following? Let's assume someone tells you, "Tomorrow, when you hear someone honk five times, go outside and I'll be there." Then, the next day, you go outside when you hear someone honk five times, even though you have never previously encountered any reinforcing consequences for doing so in that kind of situation. Or, for a slightly longer-term example, if a colleague behaves disagreeably, you may think, "The next time she acts like that, I'm going to give her a piece of my mind." Then, three weeks later when your colleague does something similar, you do exactly that. The mechanisms at work become even more interesting in the extreme long term, when we humans do things in the present that seem to be governed by how we think things are going to be after we're dead, whether it has to do with prospects of going to heaven, our children's financial future, or the desire to "finally find peace."

A more technical way of expressing the same question is to base it on Skinner's definition of verbal behavior that functions as rules or instructions. He wrote that this behavior specifies behavior and consequences (Skinner, 1966). That leaves us with the question of how a verbal behavior now can specify behavior and consequences that are not taking place in the present and that the individual has not earlier experienced. And how are we able to carry out new actions in order to achieve such consequences? How do we manage to stop smoking based on the assertion "Stop smoking, or you'll have a high risk of getting lung cancer," when the consequences we face in doing this are primarily of a punishing type? The consequences that result could include short-term effects like withdrawal symptoms or the loss of pleasant company during smoke breaks, whereas "not getting cancer" is very abstract and temporally distant. How does this work?

Skinner answered this question by referring to "a long history of verbal conditioning" (Skinner, 1957, p. 360), but he never pinpointed how to describe such a potential learning history. Remember that a fundamental principle for how both antecedents and consequences acquire their governing functions for behavior is that they are contingent on the behavior they influence. Experimental behavior analysis sees the direct contiguity between stimuli as absolutely crucial, for operant as well as respondent conditioning. Skinner distinguished between this and rule-governed behavior, maintaining that a complex learning history in one way or another bridges this dividing line. But what would such a history look like? That is a question he never answered. Early on, he mentioned the possibility of human language involving something more than the principles of operant and respondent conditioning, which he had accounted for (Skinner, 1938), but he later abandoned this alternative.

Several leading behavioral analysts, like Michael (1986), Parrott (1987), and Schlinger (1990), have brought this issue up, along with the observation that a convincing answer has long been overdue within behavior analysis.

THE COGNITIVE ANSWER AND ITS PROBLEM, AS SEEN FROM WITHIN BEHAVIOR ANALYSIS

A common answer to the above problem about how people go about relating to things and events that are not present is what could be summarized as cognitive theories or information processing theory. There are many variants on this approach, which has historical roots extending back millennia (Ellis, 1989). In their modern form, these approaches have in common the assumption that humans are equipped with some type of internal structure for carrying information. Common designations for these internal structures are "schemas" and "mental representations." These structures are influenced and transformed by external events and are credited with playing a crucial role in governing human behavior. External stimuli or events give rise to an internal processing activity in humans, and this internal processing helps explain our behavior in a way that isn't possible by means of analyzing the external events in themselves. Causes of human behavior, especially more complex human behavior, are therefore primarily sought in these internal processes.

Within behavior analysis, on the other hand, the guiding principle is that the causes of different behaviors are things that occur in the context of these behaviors. Causes are processes outside the actual behavior, and therefore they are accessible to direct influence, at least in principle. This is a pragmatic approach, adopted because it supports the aim of behavior analysis, which is not just to achieve prediction, but also to achieve influence. In this light, assumed internal structures, like schemas, are problematic. After all, such structures are not accessible to direct influence. They are merely assumed, and they are not available for contact in time and space. All we can contact is their effects: the phenomena they are assumed to cause. If something cannot be contacted in time and space, then it also cannot be influenced in a direct way. In behavior analysis, these theories based on assumed internal structures and mental representations are seen as a historical remnant of prescientific discussions that included the soul or the psyche (Skinner, 1963).

Arguments for internal structures as causes of behavior follow the same pattern as everyday expressions we use when, for example, we say that someone is eating "because she is hungry." How do we actually know that someone is hungry? This is simply a conclusion arrived at based on the person's behavior, which is what we can observe. She is acting in a certain way in relation to food. The expression "being hungry" only summarizes a number of behaviors and phenomena that we can observe or contact in some other way. Some of these phenomena can be contacted by everyone present, such as seeing the person eating food or hearing her talk about it. Other phenomena are available for contact only to the person who experiences them, like the feeling in her stomach. But the assessment "She is hungry" is nothing more than these phenomena taken together. If these phenomena are not present, then "hunger" as an internal object disappears. Of course, this way of expressing ourselves is often linguistically convenient. It is, to use one of the terms I have accounted for in this chapter, an example of how we tact our own or someone else's behavior. But the fact that "she is hungry" does not suffice as a scientific explanation for why that person is eating—at least not in behavior analysis. The expression "she is hungry" is simply a summary of the behavior one wishes to explain. In order to answer the question of why someone is eating, it is necessary to examine the context of the behavior. You have to search among the antecedents and consequences surrounding the behavior of eating. The causes must be sought both in the present and in the historical context.

Many modern cognitive theories or information processing theories use neurobiology in their explanatory models (Siegel, 1999) and see different brain structures and the activity taking place within them as causes of behavior. Although seemingly more scientific, this is much the same as the assumptions underlying hypothetical structures like schemas. And even though brain structures are obviously available for contact in time and space, the basic, pragmatic objection remains: Brain structures, or what occurs in them, are not external to the behavior they are said to cause. They are, in fact, a part of the same behavior. If I lift my hand, events are taking place in my arm, my shoulder, my aorta, my brain, and more. But all of these phenomena are a part of my action of lifting my hand. In behavior analysis, behavior is defined as an action performed by the organism as a whole, and a part of the action cannot explain the action in its entirety. The behavior of "lifting my hand" is an action performed by me as an entire organism, and what takes place in my brain cannot suffice as the cause for my action, any more than what happens in my arm, my shoulder, or my aorta. All of these are contributing elements and therefore are parts of the action. And in the behavior analytic approach, the cause of the act cannot be a component of the action itself; causes must be sought in events that precede and/or follow the action. They are to be found in the action's context. Again, this is because behavior analysis seeks causes that are accessible to direct intervention, based on our fundamental goal of influencing behavior.

This means that to a behavior analyst, cognitive theories, with their explanations based on internal structures or mental representations, are a dead-end street. (For a more thorough account of the scientific-theoretical approaches within behavior analysis, see Moore, 2008, and Wilson, 2001.) If mentally "representing" is an important part of a sequence of human behavior, the behavior analytic question would be How are we to understand this presumed human activity of "representing"? This gives us even more human behavior to explain, which is then done by performing analyses of events in the action's context. Referring to mental representations does not, as seen from the behavior analytic perspective, present us with any useful answers to questions concerning the causes of human behavior. This is, at best, only a description of more human behavior that then needs to be explained. At worst, this model places the causes of behavior in an assumed internal world that is not accessible to scientific psychological analysis. That said, neurobiological research—and its different theories about how what takes place in the brain plays a role in the actions of human beings—is of course an important field of research in its own right. But such research cannot answer the questions within behavior analysis about how we can understand and influence behavior using psychological interventions.

Cognitive theories are multifaceted, and they vary among themselves. My intention here has not been to do an exhaustive comparison between behavior analysis and any of these approaches, or to account for how such approaches could succeed in solving the behavior analyst's dilemma, explained above.

There is extensive literature on this topic available for interested readers (e.g., Andersson, 2005; Solso et al., 2005). My purpose has simply been to point out, from a behavior analytic perspective, what unites these other perspectives, and what makes them unacceptable as alternatives.

The fact that cognitive and behavior analytic approaches are very different from each other does not eliminate their points of contact in connection with the phenomena described. (I will return to this topic later in this book.) Neither do the differences imply that a dialogue between these approaches is useless. On the contrary, there are writers who argue that we are currently at a point where such a dialogue could be productive (Overskeid, 2008).

BEHAVIOR ANALYSIS AND HUMAN THINKING—A BRIEF SUMMARY

An examination, from a behavior analytic perspective, of what we normally call thoughts or thinking starts out from the fact that we do indeed speak about this phenomenon, which means that we have learned to speak in this way. This has taken place in accordance with the same principles that govern other human behavior, that is, through operant and respondent conditioning. The way we have learned to speak—aloud to begin with, and then silently—has shaped the phenomenon we are to analyze. Large parts of this phenomenon—parts that are crucial to our purpose—are, then, what Skinner called verbal behavior. This means the behavior of thinking is a part of a greater class of behavior; it is one part of verbal behavior. Verbal behavior is primarily a social action. It occurs as a result of a listener being present, and it is reinforced by the listener's actions. At the same time, this behavior can have dramatic effects on the listener's behavior. Verbal behavior can have immediate consequences when it is uttered, as when someone gives me something I have asked for. But its effects can also be greatly delayed, as when I travel abroad to visit a certain place because of something my parents told me twenty years earlier. The latter has been called rule-governed behavior in behavior analysis ever since Skinner first used the term. When we first learn this type of behavior, we are listeners and someone else is the speaker. Someone else supplies the rule, and the listener learns to act on it. But the same behavior can occur after we become both speaker (silently) and listener, within ourselves. We learn to formulate rules, which we then follow. Human behavior is often influenced by self-generated rules.

Behavior analysis has faced significant problems in explaining the effects of verbal behavior on the listener's behavior, and not only in regard to effects that are delayed in time. A verbal statement can have short-term consequences that are hard to explain based on the classic principles. The statement "Hide behind the statue; the two guys in brown jackets are out to get you" can have a rapid effect on the listener's behavior. This is true even if the listener has never had any unpleasant experiences connected with people in brown jackets, and even if she has never been assaulted or hidden behind a statue when threatened. Similar effects can follow upon self-generated rules: rules that can only be apprehended by the person following them. There have been proposals for an explanation of these effects through respondent-like associative learning of some kind (Parrott, 1984). However, these explanations seem far-fetched when it comes to more complex linguistic behavior, and they haven't led to any research-related progress connected to behavior analysis's second objective: influencing behavior (S. C. Hayes, Barnes-Holmes, & Roche, 2001).

A REDUNDANT QUESTION?

My main quest in writing this book is to describe theoretical conclusions based on new experimental findings concerning human language and cognition. I also want to show how we can use those conclusions to solve behavior analysis's dilemma concerning verbal and rule-governed behavior. And all of this is done with the purpose of increasing our understanding of complex human behavior and our ability to influence it. This knowledge can then be applied to the types of problems that make people seek psychological help, and therefore can help provide clinical tools in working for change. But before exploring this important issue, let me bring your attention to the question of whether working to understand and influence the function thinking has for humans is, in fact, clinically relevant. In chapter 3, I'll examine whether this has any significance in working for change.

CHAPTER 3

Is the Power of Thinking a Clinically Relevant Issue?

It may seem far-fetched to question whether the role of thoughts is crucial in working for psychological change. Is it not absolutely obvious that people's ways of thinking are decisive for how they act, and that if we want to influence this, thoughts and thinking need to be in focus? We assume that this is so in most of our daily discussions and attempts to influence one another. You might even say that this is also the basis for my writing what I am just now writing. Is it not my intention to influence the way you, as a reader, think?

Still, there are at least two reasons why I want to bring some attention to this issue. The first is behavior analysis's own history, including questions concerning the power of thinking. The second is an ongoing debate within empirical psychotherapy research regarding whether strategies specifically for changing ways of thinking are an essential element in therapy. I will begin with behavior analysis and its questions regarding the power of thinking.

PRIVATE EVENTS AND THEIR INTERACTION WITH OTHER BEHAVIOR

Part of the behaviorist tradition completely rejects the possibility of scientifically studying what Skinner referred to as private events. Watson, the father of behaviorism, advocated this position (Morris & Todd, 1998). To him the essential issue was that of objectivity. Anything scientific had to be observable to several independent observers. As the previous chapters have made clear, Skinner rejected this position, maintaining that just because an observation is jointly agreed upon does not necessarily make it scientific and, further, that a scientifically valid observation can be made by a single individual. According to Skinner, the scientific validity of a statement is determined by whatever controls the statement. If a person says, "I have a pain in my left leg," and the statement is controlled by stimuli that the person contacts in his left leg, then the statement is scientifically valid. It is what Skinner called a pure tact. What is being tacted in this case is no different in kind from phenomena that might be observable to more than one person, such as someone moving his left leg. More specifically, it is not the case that the former is "mental" in nature and thus different in kind from the "physical" nature of the latter. What is being tacted is a phenomenon inside the person's own body, such as pain, or the person's own behavior (Skinner, 1953). If someone says, "I thought the pain would soon go away," the individual is also tacting his own behavior—in this case, silent verbal behavior.

Behavior analysis has thus always held that we can understand private events in the same way we understand other types of behavior. The question of how this understanding impacts our work to influence behavior has not, however, been quite as clearly answered. Behavior analysis's focus on external variables to explain behavior has often led to viewing phenomena such as thoughts and feelings as a kind of by-product, without any decisive significance for working toward changing behavior. There is an intrinsic tension within behavior analysis between its affirmation of the validity of private events, on one hand, and the tendency to disregard them in practice, on the other. This tension is evident in a number of Skinner's texts (Schnaitter, 1978). One example is his assertion that the analysis of private events often does not significantly impact our efforts to influence human behavior (Skinner, 1953). But how are we then to understand his intention when writing about the importance of being aware of oneself and one's behavior in order to influence it (Skinner, 1974)? In the same way, Skinner's analysis of rule-governed behavior indicates that private events play a crucial role in controlling the rest of our behavior. Even though rule following does not originate in the private sphere, private events do seem to play a role in how this behavior evolves. This is particularly evident in behavior influenced by self-generated rules; that is, situations in which the speaker and the listener are one and the same.

Another sign of the divide in behavior analysis in this regard involves the issue of how private events can function as governing consequences. In analyses of clinical problems, it is rather common to assign governing functions to emotional conditions like anxiety and dysphoria; they may, for example, be seen as the basis for avoidance. And yet these conditions are, at least largely, private events. Leading researchers and theorists have pointed out this inconsistency (Dougher & Hackbert, 2000; Kanter, Busch, Weeks, & Landes, 2008).

Clearly, behavior analysis has long contained this unsolved dilemma concerning private events and their influence on human behavior as a whole. An important aspect here is the possible power of thinking, and if it is real, how that power functions. In many respects, this is essentially the same dilemma as the one I described in the previous chapter; that is, the question of how we are to understand and influence rule-governed behavior.

THE DOMINATING POSITION OF COGNITIVE THERAPY MODELS

In clinical psychology and psychotherapy outside the behavior analytic tradition, the power of thinking is center stage, and this has been the case at least since the 1970s. The rise of cognitive therapy and its triumphal progress throughout the world of psychotherapy have brought forth an outlook frequently summed up in a well-known quote from the Greek philosopher Epictetus, and cited also by Aaron Beck, the most prominent figure in cognitive therapy: "Men are not moved by things but by the views which they take of them" (Beck, 1976, p. 47). This outlook largely corresponds with how most of us view the circumstances of our everyday lives, assuming that the way we think about things influences our other behavior a great deal.

Cognitive therapy is a therapy package with several different components and a variety of forms. At the center of the model, however, we find the human being as an information-processing entity. What is crucial is the power of thinking. The fundamental strategy in therapy is therefore to help the client reconsider the very way he regards what happens to him. The task of the therapist is, through reasoning and different types of experiments, to help the client reach a more adequate notion of himself and his situation and thereby enable him to act in a more adequate way. Cognitive distortions and misinterpretations of processes and events are seen as a very important part of people's psychological problems, and therefore a more realistic way of thinking is seen as the solution. Work in therapy is dominated by attempts to correct distorted interpretations of reality. This approach is clearly described in Beck's early texts on cognitive therapy (1963, 1964, 1967), and it remains essentially intact in more recent texts, both by Beck himself (1991) and by other prominent cognitive therapists (e.g., Freeman, Pretzer, Flemming, & Simon, 2004; Leahy & Dowd, 2002; Salkovskis, 1996).

Cognitive therapy models have demonstrated favorable treatment efficiency for the most common clinical diagnoses (depression and anxiety) as well as for a number of other conditions (Beck, 2005; Dobson, 1989). This seems to support the idea that the question of whether it is important in psychological therapy to change thoughts and ways of thinking is unnecessary. We all have the impression that our way of thinking plays a crucial role in how we act, and most of the well-substantiated forms of psychological therapy basically agree on this fundamental outlook. Still, clinical psychology as a whole contains a tension and a dilemma that actually corresponds to that of behavior analysis. While behavior analysis has a history of skepticism with regard to the power of thinking, the dominating cognitive therapy models have a history of confidence. The basis of this confidence has nevertheless been called into question these last few years.

COGNITIVE THERAPY MODELS ALSO STRUGGLE WITH THE POWER OF THINKING

The fact that a certain therapeutic model shows positive effects does not necessarily mean the theory underlying the model offers the correct explanation of these effects. Let's say I recommend a specific vegetarian diet based on the conviction that meat-eating is what brings about an increase in weight. Let's also say this diet leads to a reduction in weight. This does not necessarily mean that my explanation is correct. The diet's effect could be explained by factors other than its being vegetarian. Comparing this example with modern psychological therapies and theories about how they achieve their effects is, of course, a simplification, yet there are similarities.

How do we know that the core assumption of the cognitive therapy model, about the fundamental power of thinking, is correct? We know that a number of cognitive therapy approaches, as whole packages, demonstrate favorable effects. But why is this so, and to what extent is it due to the core assumption? Cognitive therapy is a composite therapy consisting of many different components. In addition to techniques meant to influence people's ways of thinking, it includes, for example, behavioral activation and encouragement to approach previously avoided situations (traditionally called "exposure" in behavior therapy). Maybe these features have at least as much to do with the efficacy of the therapeutic packages as the assumptions concerning the power of thought.

In recent years, several different studies have attempted to answer these types of questions by distinguishing the different components in cognitive therapy and comparing their effects separately. The researchers who have felt compelled to question the theoretical side of the cognitive model have tried to separate the techniques in therapy that are aimed at influencing a person's way of thinking (cognitive techniques) from the behavior therapy techniques that are a part of the cognitive package, such as behavioral activation and exposure. A well-known study compared depression therapies in this way (Jacobson et al., 1996). A group of depressed patients (150 altogether) was divided into three groups. All of the patients were given therapies that consisted of twenty sessions. One group was given only the behavioral activation component of cognitive therapy; all therapy strategies aimed at influencing their ways of thinking were excluded. The second group was taught ways of finding alternatives for negative automatic thoughts, in addition to receiving the behavioral activation component. The third group was given complete cognitive therapy, including work aimed at correcting underlying beliefs. The results showed the same level of treatment efficiency in all three groups. The study included independent observers to ensure that the different therapies were delivered in accordance with the criteria for each group. Follow-ups were carried out after one and two years and still showed no differences between the groups (Gortner, Gollan, Dobson, & Jacobson, 1998). A subsequent study has since returned similar results (Dimidjian et al., 2006). In this later study, the behavior-based therapy model was modified to make it more consistent with established behavior analytic principles pertaining to depression (as outlined in Ferster, 1973). For the more serious depressive conditions, this study actually showed a more positive effect from the behavior-based intervention model than the cognitive model. For less severe depressive conditions, the effects were equivalent.

Anxiety problems have also been the focus of studies in which therapeutic packages including versus excluding specific cognitive techniques were compared. The results have been mixed (Marks & Dar, 2000; Öst, Thulin, & Ramnerö, 2004). Most studies have not pointed to any significant differences. However, some studies of social phobias have shown more favorable results following therapy based on a consistent cognitive model compared with therapy that didn't include the cognitive element (Clark et al., 2006). The alternative therapy model in this study was classical exposure.

It is a very delicate and difficult task to try to discuss the results of individual therapy studies, how the studies were carried out, what might explain the differences, and so on. Researchers often have differing opinions about how to interpret the details and what conclusions to draw. Individual studies of this kind cannot be held to be conclusive one way or the other. A recently published review did, however, reach the conclusion that "there is little empirical support for the role of cognitive change as causal in the symptomatic improvements achieved in CBT" (Longmore & Worrel, 2007). Even leading researchers who themselves clearly adhere to the cognitive model say that the research situation today is such that we cannot make any decisive conclusions about what exactly mediates the changes in cognitive therapy (Hofmann & Asmundson, 2008). And this is said about a therapy model that has dominated the scene for more than thirty years! During this same period, results within the empirically supported therapies more generally haven't seemed to improve as cognitive models have come to predominate (Öst, 2008).

This means that even outside of behavior analysis, the question concerning the power of thinking is in many ways unanswered. There is still no clear scientific support for the notion that a specific focus on thoughts and the power of thinking is essential to clinical therapy work. Such a focus is definitely a part of existing, functioning therapies. But we do not know what the significance of such a therapeutic strategy is, and even if an emphasis on thoughts and thinking was proven to be essential, we still wouldn't know how this focus should best be implemented in practice.

THE NEED TO PURSUE THE ISSUE OF THE POWER OF THINKING

Clearly, behavior analysis and cognitive therapy models have an issue in common to struggle with. How are we to understand verbal behavior or cognitive processes, and how could a potential understanding in this area be used in therapy? The two traditions approach this question from very different starting points. Behavior analysis has a history of having in large part neglected the question, at least in practice. The different cognitive models have a history of confidence that now seems premature. I believe that there are two main considerations that call for a renewed attempt to answer this question.

First, there is a great deal of knowledge, besides clinical therapy studies, to back up the power of thinking and the essential function it has for humans. It is very difficult to seriously argue that people's ability to imagine things, make plans, solve problems "in their minds," and so on is something irrelevant. If it were irrelevant, why would we spend so much time and effort engaged in these types of activities? In addition, if we view them as irrelevant, then how can we explain the phenomenon of rule-governed behavior, discussed in chapter 2? How do people go about relating to things and events that are not present?

Is it possible that these phenomena could be important to people in other areas in their lives, but that they might be irrelevant to psychological problems and what we call psychopathology? That seems unlikely, and difficult to defend, as well. Our knowledge of different kinds of psychological problems indicates that what we commonly call thoughts and thinking often do have an impact. We know, for example, that people who are depressed have a typical way of recalling memories (Williams, 1992). We also know that verbal behavior tends to affect mood, and a depressed mood is, by definition, a part of depressive conditions (Dougher & Hackbert, 2000). We find characteristic patterns of thinking, and what within cognitive science is called cognitive bias, in connection with other psychological problems, as well (Wray, Freund, & Dougher, 2009). Is it reasonable to argue that all of these observations are irrelevant and that thinking is nevertheless still only some form of secondary phenomenon that lacks significance for the type of influence that psychological therapy seeks to achieve? Is it not more reasonable to assume that we do not yet know enough about the way these phenomena operate for us to achieve this influence? Perhaps our knowledge is simply too limited and in need of greater depth.

The second argument for intensifying our research efforts in this area can be found in precisely those studies that question this focus on cognitive phenomena in therapy. The reason is that even if we exclude specific cognitive techniques from the therapy protocol, this does not mean we are excluding verbal behavior as a whole. This is, of course, obvious. Any psychological therapy where a therapist meets with the client to work with the client's difficulties is accomplished through conversation, by talking. And even when something like a concrete phobia, say a spider phobia, is treated using exposure, a large part of the therapy consists of talking about the different steps of the treatment, when they will take place, and what will happen when they do. All types of psychotherapy, regardless of the underlying model, are largely talk therapy.

Let's take a closer look at one of the models that, in more recent years, could be seen as a good example of a therapy that excludes cognitive components; namely, behavioral activation, which is used in treating depression (Martell, Addis, & Jacobson, 2001). In behavioral activation, the therapist refrains from deepening conversation that deals with the content of thoughts and does not enter into dialogue to question the client's way of thinking. There is a complete absence of attempts to map out underlying beliefs in order to replace them with other beliefs. Instead, the therapist puts a lot of effort into explaining to the client both how to understand depression as a condition and how the therapy is expected to work. This therapy employs a specific questioning technique that is aimed at helping the client observe and describe his own behavior. In the terminology used earlier, the client learns to tact his own behavior. Homework assignments are planned, and the therapist encourages the client to describe his experiences afterward. In short, the therapy is, of course, full of verbal behavior, both in the form of actions by the therapist and in the form of skills that are encouraged in the client. Indeed, no advocate of this therapeutic model would maintain anything else.

A STRATEGIC STEP BACK

Taken together, the issues discussed in this chapter mean that we have cause to deepen and develop our analysis of verbal behavior in hopes of being able to develop better therapy models. Could it be that a better understanding of how we humans influence each other and ourselves through verbal behavior, whether silent or spoken aloud, is exactly what is needed? This book is written from the stance that the answer is yes, and out of a conviction that, in recent years, basic research in behavior analysis has provided new findings that can contribute to this end. To reach this understanding, it is necessary to take a strategic step back from research that focuses on therapy models and comparisons between them and instead conduct basic experimental research that attempts to clarify how verbal behavior is best described and how its role in controlling human behavior is best understood. This research may help answer such questions as "What is a cognition?" and "How do we manage to say something to someone (or to ourselves) in a way that influences their (or our own) behavior?" and "How are these phenomena connected to psychological problems?"

In part 2 of this book, I'll take this step back to basic research and its conclusions in order to find the answers a modern behavior analysis of human language and cognition can give to these questions. The descriptions are brought together under the concept of relational frame theory, and describing RFT is the main quest of this book. Simply put, I maintain that to understand, predict, and influence human behavior, we need a theory of human language based on experimental research. RFT is such a theory.

PART 2

Relational Learning

CHAPTER 4

Derived Relational Responding as the Fundamental Element in Human Language

If the human behavior underlying language and cognition is different from other human behavior in any significant way, in what way might it be different? It has always been important within behavior analysis to define the concepts that are used, such as operant and respondent learning, in an exact manner. Anyone who maintains that verbal behavior deviates from other behavior must be able to describe exactly how it does so.

In my attempt to outline an analysis of verbal behavior using relational frame theory, I will begin an exploration of three questions:

- What kind of behavior do we refer to when we speak of "verbal behavior"?
- How does this behavior interact with human behavior as a whole?
- What are the factors that govern verbal behavior?

In order to answer the first question, we must describe verbal behavior somehow. But where should we begin? The behavior we loosely refer to as verbal appears to be common, and behavior analysts have attempted to pinpoint its core phenomena by mapping out linguistic interactions; however, earlier efforts didn't result in any decisive progress (S. C. Hayes, Barnes-Holmes, & Roche, 2001, pp. 10-11). The field of research that opened a gateway to something new was work carried out by Murray Sidman and colleagues to help children with specific difficulties in learning language (Sidman, 1971; Sidman & Cresson, 1973). Sidman described how children with learning difficulties, while practicing their reading skills, could learn to relate stimuli to each other in a way that was hard to understand based on known principles of operant and respondent conditioning. The study participants related certain stimuli even though the experiments did not establish contingencies between the stimuli in ways thought to have been necessary for those types of relations to occur, according to contemporary thinking on the principles of operant and respondent learning.

DERIVED STIMULUS RELATIONS

Human language shows a very high degree of complexity. Critics of behavior analysis have maintained that this complexity cannot be explained by operant and respondent learning and have challenged behavior analytic researchers to show how behavior of such complexity can be learned; for example, how new linguistic statements can be produced without having been specifically reinforced in the individual's history (Chomsky, 1959, 2006). It has long been known that relations between stimuli can be established in ways that are hard to understand based on the fundamental principles of learning theory, but the phenomenon was not clearly described until Sidman's experiments at the beginning of the 1970s (Sidman & Tailby, 1982). Let's take a closer look at this phenomenon, known as *derived stimulus relations,* that is, relations between stimuli that appear without having been learned or trained specifically. Since Sidman's first experiments, similar experiments have been performed many times and have confirmed and further developed his original findings.

Here is a description of the procedure for a typical experiment of this type. A group of study participants are taught to choose a certain arbitrary stimulus, which we can call D, when they are presented with another arbitrary stimulus, which we can call E. All of the stimuli used in the experiments are meaningless to the participants at the outset. They usually consist of figures with no meaning or letters in a nonsense combination.

Examples of typical stimuli used in experiments demonstrating derived stimulus relations

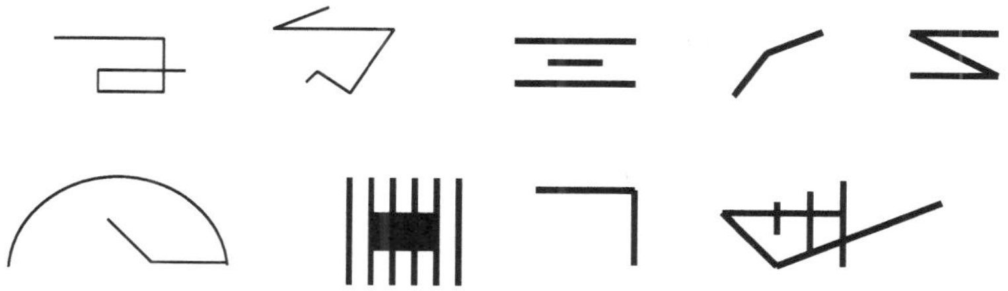

URD **OXQ** **TGG** **GCF**

EWT **RKO** **AFD** **HFU**

The participant is shown E and given the option of choosing either D or another stimulus, which we can call F. If D is chosen, this is reinforced. This is repeated with different options involved, but choosing D when E is presented is consistently reinforced. Thus far, E → D has been specifically trained via operant learning. E now functions as a discriminative stimulus for choosing D. If E is shown with D as a possible option, the probability is high that the participant will choose D. This procedure is called matching to sample. In this case, E is the sample and the task is to match another of several possible stimuli (D or F).

Let's move on to the phenomenon we want to take a closer look at, wherein new relations arise without being specifically trained: The conditions are now changed by presenting D, after which participants can choose between several different meaningless stimuli, one of which is E. In this case, participants are likely to choose E, even though D hasn't previously been presented as sample and there is no learning history where choosing E has been reinforced. The latter is ensured by using only meaningless stimuli in the experiment; they are made up purely for the experiment and there is thus no prior relation among them. After training E → D, another relation has been established between the stimuli without specific training: D → E. This is known as a *derived stimulus relation,* as it is not trained directly but is derived by the study participants in a situation of this type. We train one relation (E → D) between stimuli, and another one (D → E) is *entailed.*

A typical experiment

Trained with feedback

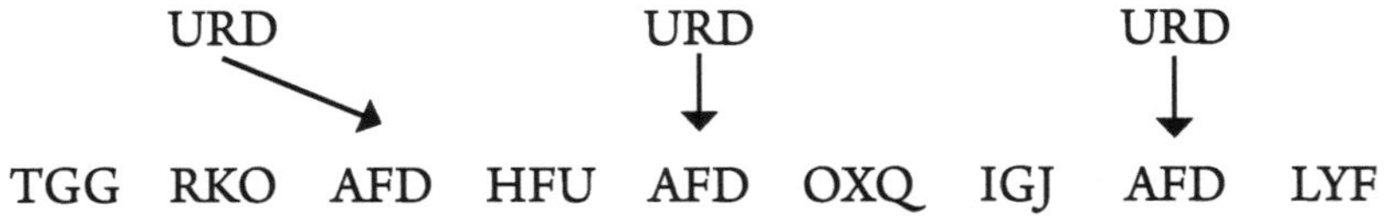

Tested without feedback

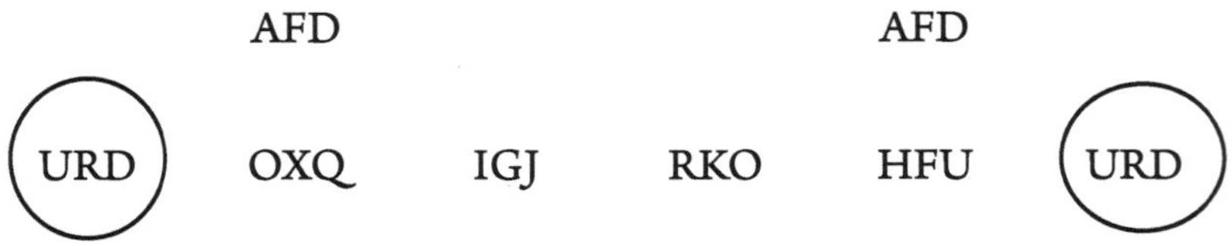

Most people may feel that this is self-evident. But the fact that it appears self-evident is probably due to the fact that this is something we are constantly doing. This behavior is universal among humans, which makes it seem natural to us. But if we are to explain verbal behavior, one of the difficulties lies precisely in that what we want to explain is something natural or obvious to us. And yet, how obvious *is* this from a larger perspective? From repeated studies involving different animal species, we know that this ability has not been convincingly shown in any other species[1] (S. C. Hayes, 1989), not even

1 The fact that there has been no demonstration of other animal species showing derived stimulus relations does not mean we should view this skill as something exclusively human. Future studies, perhaps of higher quality, could change our understanding in this respect. It seems clear, however, that if other species do have some degree of this ability, it is to a much lesser extent than humans do.

chimpanzees that have had many years of "language" training with humans (Dugdale & Lowe, 2000). Humans, in contrast, show this behavior from at least two years of age (Devany, Hayes, & Nelson, 1986).

That a trained relation (E → D) entails an additional relation (D → E) is just one aspect of derived stimulus relations. It is possible for two or more trained relations to lead to the emergence of one or more other relations based on a combination of the trained relations. Let's say that we perform training of another relation: If D is present, choose Z, and do not choose any other available stimuli. This, through direct training, provides us with the relation D → Z. If this is trained, an additional relation will be derived: Z → D. This takes place in the same way that the relation between D and E occurred. We have trained two relations, and two more have been derived: E → D and D → Z have been trained, and D → E and Z → D have been derived. But since D is a part of both relations, these two can be combined. Now, if Z is presented as an initial stimulus and E is a possible option, the likelihood is high that study participants will choose E, provided that no other stimuli that are trained directly or that are in a derived relation to Z are among the possible options. The same goes for E as an initial stimulus with Z as one of the possible options. And this is despite the fact that E and Z are not in any directly trained mutual relation. The relations that have been trained display *combinatorial mutual entailment.*[2] Although participants were trained in only two relations, they derived four other relations.

---

2 The terms "mutual entailment" and "combinatorial mutual entailment" for the phenomena described are used here as in the theory described in this book: relational frame theory (RFT). However, other terms for these phenomena can be found in the literature. Sidman uses the term "symmetry" for the former and "transitivity" for the latter. The term "stimulus equivalence" is often used for the phenomenon as a whole. Regarding combinatorial mutual entailment, this term is often shortened to "combinatorial entailment"; however, I will consistently use the full term throughout, as it more clearly shows that we are referring to mutual entailment that is combined.

Mutual entailment and combinatorial mutual entailment

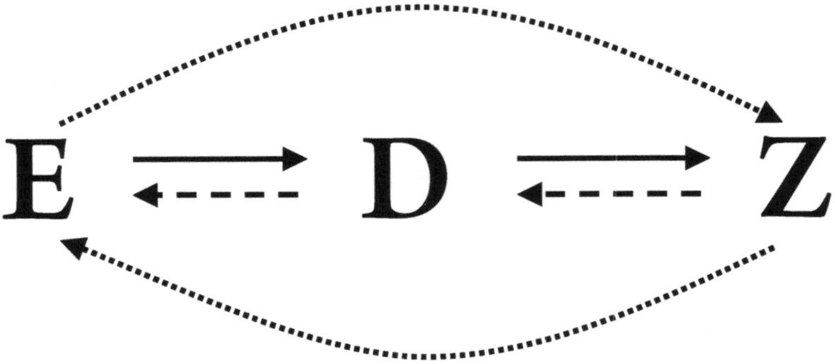

Directly trained

Mutual entailment

Combinatorial mutual entailment

The experiments I have described may seem rather abstract. But they show us that humans can do something that other animals either cannot do at all or possibly can do, but only to a much lesser extent. These experiments also demonstrate a type of learning that does not seem to follow the principles we know as operant and respondent learning. In these experiments, the training is aimed at making E function as a discriminative stimulus for choosing D. However, D also acquires a discriminative function for choosing E. How does this happen? In addition, E and Z each acquire discriminative functions for choosing the other without the two being in any directly trained relation with each other. In this case, they acquire their functions via combination of the functions trained to occur between E and D (if E is given, choose D) and between D and Z (if D is given, choose Z).

If we were to carry out such experiments with animals other than human beings, the animal (say, a dog or a monkey) will act in accordance with the trained relations after receiving the same type of training. There is a high probability that the dog will choose D if E is presented, and Z if D is presented. But as far as the remaining options are concerned, it will simply choose randomly. There do not seem to be any derived stimulus relations present to the animal. The same thing is true for most children up to almost two years of age.

All of this is strikingly similar to what we know from daily life. Let's say that we are training a clever parrot and a likewise clever three- or four-year-old girl, Sue, to say "pretty Carla" just after Carla steps into the room. We do this by rewarding them when they repeat after us. The parrot is given a couple of peanuts when it produces a sequence of sounds that gradually approximates "pretty Carla." Sue receives encouragement, probably by us saying, "That's right, 'pretty Carla.' That's good, Sue. Keep it up!" This type of interpersonal contact (often called tuning or mirroring) has a reinforcing effect on Sue's behavior but little to no effect on the parrot's. In this way, we train both of them to say the desired words after Carla has stepped into the room. If this is a particular type of parrot, it may sound rather like what Sue sounds like. We say that the parrot has learned to speak. Still, it is easy to show that Sue has learned some things that the parrot has not. If, at this point, we were to say "pretty Carla" when Carla cannot be seen, nothing we know about parrots or their way of behaving indicates that this has any meaning to the parrot. Sue, on the other hand, is likely to turn around, look toward the door, and so on. Neither the parrot nor Sue has been trained in accordance with "If you hear 'pretty Carla,' look for or expect Carla." The relation that was trained was seeing Carla → produce the sound "pretty Carla." However, it is likely that Sue will also act on the derived relation: the sound "pretty Carla" → Carla. This relation does not seem to exist for the parrot.

I could give many examples illustrating the ability of young humans to derive relations that don't seem to exist for other animals. We can teach a dog to react to the word "cookie" by repeatedly saying "cookie" just before we give the dog a real cookie. After the dog has had this repeated experience of the word "cookie" → real cookie, we can expect to find that when we say "cookie," the dog will in various ways behave as though a real cookie was on its way. If instead we train the dog by first giving it the cookie, and then uttering the word "cookie" after it has eaten it, the dog will never learn to react to the word, no matter how many times we repeat this procedure. The dog's learning process follows a certain order. From the relation real cookie → word "cookie" (which the dog experienced in the latter example), the dog will not derive the relation word "cookie" → real cookie. If we were to do the same thing with a four-year-old child, the child's ability to react to a derived relation will be evident. The child will react to the word "cookie," even though the relation word "cookie" → real cookie has never been trained in a direct way. The word "cookie" has acquired its function by way of mutual entailment.

On the surface, it looks like we can teach parrots and dogs parts of human language. What small children learn at an early age, however, is an additional form of responding. They are learning the fundamental skills in human languaging.

Derived Stimulus Relations as a Fundamental Process in Human Language

Researchers have long supposed that this particular way of relating is essential to human language. To illustrate the high probability of this supposition being correct, imagine the following learning situation, illustrated with a fictitious experiment. (Although it is fictitious, similar experiments have been carried out a number of times.) At seven years old, Kyle has learned to read, but he has never been in contact with any language other than English. Now Kyle gets to take part in an experiment along the lines described above, but using real words. We show the word "chair," and Kyle gets to choose between several five-letter nonsense words, one of which is "silla" (Spanish for "chair"). The word "silla" is, of course, not a nonsense word to someone who knows Spanish, but because Kyle has never been in contact with any language other than English, it is just as void of meaning as any of the other optional "words" he is presented with (maybe "chorp," "tvari," and so on). So Kyle makes guesses. If he chooses "tvari," the response is "Wrong." If he chooses "silla," which he first only does at random, this is reinforced ("Right. That's good, Kyle!"). If this is repeated until he chooses "silla" instead of other words when "chair" is presented, then Kyle learns that "chair" → "silla." As anyone who has interacted with children who are increasing their vocabulary knows, Kyle will also derive the relation "silla" → "chair" without having to be specifically trained to do this. And if he is trained in the same way to choose "stol" (Swedish for "chair") when "silla" is shown, from this trained relation "silla" → "stol," he will derive "stol" → "silla," as well. But he will also be likely to choose "chair" when "stol" is shown, and "stol" when "chair" is shown. In other words, he is acting in accordance with derived relations, and his behavior demonstrates both mutual entailment and combinatorial mutual entailment.

Mutual entailment and combinatorial mutual entailment in Kyle's language training

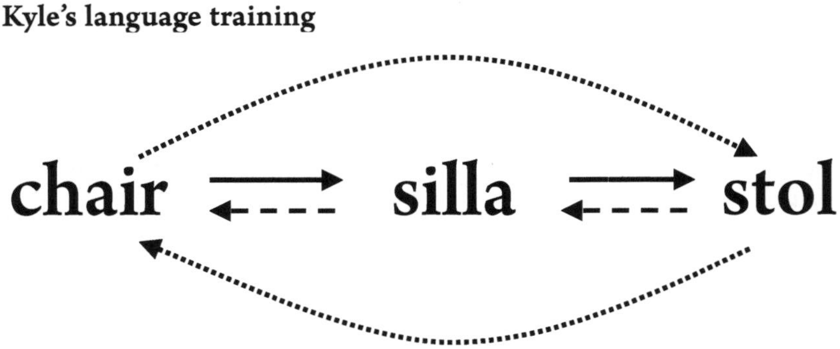

Directly trained

Mutual entailment

Combinatorial mutual entailment

Does this not seem very much like what we can easily observe in the ordinary language training that constantly goes on in the dialogue between children and their parents or other caregivers? Relations between words—both spoken and written—and different types of objects, pictures, and events are naturally trained in this very way in our social environments, and it seems obvious that all of these different relations are not specifically and separately trained.

The fact that the phenomenon of derived stimulus relations, which can be demonstrated in monitored experiments, has such a strong similarity to what we observe in human languaging skills has led researchers to the assumption that this is a fundamental process of human language. Other findings that convincingly point in this direction are from studies of individuals with varying degrees of language difficulty. These studies have shown a correlation between proficiency in language and the ability to demonstrate derived relational responding in laboratory experiments. Individuals with very low language scores tend to fail tests of derived relational responding (Devany et al., 1986).

The link between language and derived relations has also been supported by a number of studies using neurophysiological methods of measurement. In these experiments, when people derive relations in the way described above, the same type of brain activity is observed as can be recorded during obvious language-related activities (D. Barnes-Holmes, Staunton, et al., 2004; Dickins et al., 2001; Horne & Lowe, 1996).

Derived Relational Responding as Learned Behavior

Exhibiting derived stimulus relations is a form of behavior, a type of responding. But what kind of behavior is it, and how does it occur? One possible answer is that this type of behavior is there from the start. In that case, it is not something we learn to do. Being able to learn associations (respondent conditioning) or being governed by consequences of one's actions (operant conditioning) are not things a person learns. The way in which specific operants and respondents develop and contribute to a person's actions depends on that person's learning history, but the capacity to learn in these ways is present at the very outset. It is a result of evolution. However, the evidence suggests that derived relational responding is not a prewired capacity, like respondent or operant conditioning, but that it is learned, and further, that even if respondent learning is also in play (Rehfeldt & Hayes, 1998), derived relational responding is something we mainly learn through operant conditioning. What is there to suggest this?

1. Operant behavior evolves gradually, over time. Studies that elucidate the development of derived stimulus responding in young children demonstrate such gradual learning (Lipkens, Hayes, & Hayes, 1993; Luciano, Gómez, & Rodríguez, 2007).

2. Operant behavior is flexible and can be influenced. This also characterizes derived relational responding, both in learning the repertoire in itself (Lipkens et al., 1993) and in learning new, individual responses (Healy, Barnes-Holmes, & Smeets, 2000).

3. Operant behavior is influenced by conditions that precede the behavior—what we call antecedents. This applies to derived relational responding as well (Dymond & Barnes, 1995; Roche, Barnes-Holmes, Smeets, Barnes-Holmes, & McGeady, 2000; Dougher, Hamilton, Fink, & Harrington, 2007).

4. Operant behavior is influenced by its own consequences; in fact, this is the defining feature of this form of behavior. Our knowledge about derived relational responses conforms with this (Y. Barnes-Holmes, Barnes-Holmes, Smeets, Strand, & Friman, 2004; Y. Barnes-Holmes, Barnes-Holmes, & Smeets, 2004; Heagle & Rehfeldt, 2006; Berens & Hayes, 2007).

This means we can formulate an answer to the first question I posed at the beginning of this chapter: Within RFT, what kind of behavior do we refer to when we speak of "verbal behavior"? *Verbal behavior is the behavior of relating stimuli or events in a particular way.* As certain relations are directly trained according to the principles of operant and respondent learning, the verbally competent human being derives additional relations—relations that need not be trained directly. The ability to relate stimuli in this way is in itself a learned ability, learned through operant conditioning.

If this is correct—that is, if the fundamental process in human language can be understood as operant behavior—then interesting and important possibilities arise. Operant behavior can be influenced. An understanding of this process would mean there are prospects for influencing human behavior in all situations where "languaging" is a contributing factor. If so, this applies to a very broad field. For example, we could help children with difficulties connected with learning language, we could improve pedagogies in general, and we could help people when their way of thinking creates problems.

In order to approach this question—how influencing of derived relational responding can occur (for example, in psychological therapy)—we first have to improve our understanding of how these responses, and thereby human language, interact with and influence other human behavior. We also need to understand what controls, or governs, this kind of relating. So far, I have simply described a phenomenon that could be interesting. But wherein lies its power to affect our actions? This is the second question I mentioned at the beginning of this chapter. To answer it, we first need to take a closer look at the concept of stimulus function.

STIMULUS FUNCTIONS AND THEIR ALTERATION THROUGH RESPONDENT AND OPERANT LEARNING

A certain stimulus or event has a function for a certain organism's behavior when the behavior occurs in relation to, or under the influence of, this stimulus. When I see a car, this car has a function for my vision; it influences the fact that I see what I see. Then, if I start walking toward the car to get closer to it, the car has an additional function for my behavior: it influences me to approach it. But the car obviously does not have this function regardless of other factors. The same car can actually have a function of avoidance for my behavior if connected with a different set of factors. For example, if it is moving toward me at a certain speed, it can make me step aside when I see it. These are known as *stimulus functions*. What kind of function a certain event has in relation to my behavior can be understood only through analysis of a specific situation. Here is a simple example: The light is turned on when I am in a dark room. This can have different stimulus functions for me, and only a closer account of the situation in question, and my responses to it, can help clarify what these stimulus functions are. The light can function as an unconditioned stimulus that triggers my eyes to blink. It can also have a punishing function in relation to the behavior I was engaged in, so that I cease my activity when the light is turned on. It can function as an establishing operation, under which something else I can then perceive becomes a discriminative stimulus for drawing closer, like if I were to approach a cupboard that I didn't see when the light was off.

My point here is to emphasize that the function of a stimulus is not an inherently given quality of the stimulus. Its function can be determined only through an analysis of the wider situation (the context) and the individual's response. The same stimulus can have different stimulus functions. It can affect the behavior of an organism in different ways. The effect depends first on which organism is involved, as this specific organism's learning history is included, and second on the context in which the stimulus occurs. When I see a red car in front of me, what I see is of course to a high degree governed by this car's characteristics. At the same time, we all know that seeing the words "a red car" can make at least some readers visualize a red car, although no such car is present. The probability of this "seeing in the absence of the thing seen" may be increased if another word is added: "Imagine a red car." This phrase has a stimulus function that in some ways is similar to the stimulus function of the actual presence of a red car. The words hardly have this function inherently; they have acquired such a function through the individual's learning history. In this example, our different histories are typically similar enough that most of us are able to visualize a red car in response to the phrase "Imagine a red car." The consequences of our learning histories are present in the moment, and in the context described the words have this stimulus function for most of us.

Some stimulus functions are more directly dependent on the nature of the objects involved. This applies to what we call perceptual phenomena, such as when we see or hear things. It also applies to the phenomena that within behavior analysis are known as *primary reinforcers*. These are stimuli that already function as reinforcing at the outset of an organism's life. Examples of primary reinforcers are water when we are thirsty, and food when we are hungry. But even these phenomena do not always have the same stimulus function. Contextual factors can alter their function. Something that in most circumstances functions as a primary reinforcer—like food when we are hungry—can, in certain contexts, have a punishing stimulus function. A dieter who has just eaten something she regrets might avoid taking another look at a plate where there is more of the same tempting food, despite the fact that she is still hungry. And in a given situation for a certain individual, perceptual stimulus functions can be completely different from what we would generally expect; for example, in a certain context, a hunter may see a moose in front of him, while what is actually there is a mushroom picker.

In chapter 1, I described the fundamental principles for learning—operant and respondent conditioning—and it is precisely these principles that describe how stimulus functions are altered. Take the example of being assaulted in a town square, and how subsequently the square itself can trigger fear. This is an example of how a stimulus, in this case the town square, acquired a new function by way of respondent conditioning. This occurs because the stimulus (the square, or a certain part of it) was associated with an event that provoked an unconditioned fear response. In this case, the assault is the unconditioned stimulus. In operant learning, events also acquire new stimulus functions based on contiguity with other events. In chapter 1, I explained how a certain facial expression in another person had a discriminative function for my behavior in addressing the person in question. How did that specific facial expression in that type of situation acquire this stimulus function? This occurred as a result of a process in which certain facial expressions of other people (A) preceded my action of addressing them (B), and certain consequences (C) resulted from my action.

Clearly, proximity of events in time and space is crucial to learning for both operant and respondent conditioning. Another way of putting this is that in both types of learning, direct relations between stimuli play a crucial role in what stimulus functions are established. Certain stimuli have either a reinforcing or a punishing function for my behavior based on the direct relations between these stimuli in my history. Certain stimuli have a function of triggering anxiety in me at a certain point, based on their direct relations to other stimuli in my history. If someone raises her voice, this may provoke anxiety or fear, perhaps because of experiences I had with my parents or with a teacher in my early years at school. The direct relation that has existed between events influences which functions are established. Behavior analysis uses the term "contingencies" for these relations in time and space (see chapter 1). The influence these contingencies have on stimulus functions is applicable both to respondent functions and to various functions in operant behavior: discriminative, motivational, reinforcing, and punishing.

An additional factor plays a role in establishing stimulus functions: what I described earlier in connection with the term "generalization." Events or stimuli are related in terms of physical properties. The facial expression that now has a discriminative function for my addressing someone does not have to be exactly the same as what I've encountered earlier; it only has to be similar enough. To enter an unfamiliar town square can cause fear if it is similar enough to the square where I was assaulted, even if it differs in some respects.

DERIVED STIMULUS RESPONSES AND ALTERATION OF STIMULUS FUNCTIONS

Let's return to the experiments discussed earlier, used to demonstrate derived relational responding. In the experiment involving Kyle, operant conditioning was used to establish relations between stimuli so that the word "chair" took on a discriminative function for Kyle choosing the word "silla." In the same way, "silla" acquired a discriminative function for choosing "stol." But, at the same time, the word "silla" acquired a discriminative function for choosing the word "chair" (through mutual entailment) and "chair" acquired a discriminative function for choosing "stol" (through combinatorial mutual entailment). This was established even though Kyle had no operant or respondent learning history that could account for how these words took on these functions. Kyle encountered contingencies of reinforcement that explain whatever was trained directly. But he also derived other relations, and these derived relations seem to govern the establishment of new stimulus functions. If altering stimulus functions is dependent on contingencies being established in the way described for operant and respondent conditioning, each and every connection must be trained directly or be established through generalization. But derived relations seem to establish stimulus functions without any such contingencies between stimuli.

The experiments I've described thus far have demonstrated that discriminative stimulus functions can arise through derived relations, but research has shown that a number of other stimulus functions can be established in the same way. One example is an experiment demonstrating that respondent functions can also be established through derived stimulus relations (Dougher, Augustson, Markham, Greenway, & Wulfert, 1994). This is especially relevant from a clinical perspective because respondent functions like fear and other emotional reactions often play a central role in clinical problems.

This experiment was performed using the type of abstract, meaningless visual stimuli described earlier. The individuals who took part in the study were first trained as outlined at the beginning of this chapter. Three stimuli in the training sessions—let's call them B, C, and D—were used in a way that put them in relation solely through mutual entailment and/or combinatorial mutual entailment. No direct connections (contingencies) according to the principles for operant or respondent learning were established among these three stimuli. The stimuli were displayed on a computer screen. Another group of stimuli—let's call them F, G, and H—were related to each other in a corresponding way, but not related to the first group. Following this, one of the stimuli in the first group (B) was given a respondent function. This was done by repeatedly administering a light electrical shock whenever B was displayed on the screen. Through this respondent conditioning, B acquired a new stimulus function. This was gauged through skin conductance, indicating what we might refer to in everyday language as mild discomfort or fear.

Parallel with this, participants were given a simple task to perform on the computer, for which they could earn a small amount of money. The purpose of this task was simply to motivate the participants to stay busy and alert in front of the screen. As the participants worked, randomly selected stimuli (B, C, D, F, G, or H) were displayed on the screen. That B caused increased skin conductance would be expected, based on respondent conditioning. However, in this experiment, C and D also caused increased skin conductance, even though neither stimulus had ever been followed by an electrical shock, nor had they been related to B based on contiguity or formal (physical) properties. However, when F, G, or H was displayed on the screen, participants did not show the corresponding increase in skin conductance.

In a second experiment with a new group of participants, derived relations were once again established among a group of previously unrelated stimuli: B, C, and D. Then, in this experiment, all of the stimuli were given a respondent function through a mild electrical shock that followed upon display of the different stimuli on the screen. After this, an extinction contingency was established for one of the stimuli: B was now displayed without being followed by a shock. As this was repeated, participants showed a decreased level of skin conductance when B was displayed on the screen, something that would be expected as a result of the extinction of a respondent function. At the same time, the experiment showed that this extinction also took place for the stimuli that were related to B through mutual entailment and combinatorial mutual entailment (C and D). The experiment also included a control group. They were not given the training that would result in the stimuli being put in a derived relation, but they were trained for conditioning by receiving an electrical shock after all three stimuli (B, C, and D) and for extinction with B in the same way as the first group. In these individuals, extinction took place in connection with B, but not in connection with C or D.

The researchers concluded that the experiments describe a process whereby stimuli can acquire and lose a function of a respondent nature without the long known conditions for respondent learning being in effect. In other words, stimulus functions were altered through derived relations. In RFT, alteration of stimulus functions based on derived relations is called *transformation of functions.*

Other, similar experiments have shown the same type of transformation of other stimulus functions, including reinforcing (S. C. Hayes, Kohlenberg, & Hayes, 1991), self-discriminating (Dymond & Barnes, 1994), mood (Y. Barnes-Holmes, Barnes-Holmes, Smeets, & Luciano, 2004) and sexual functions (Roche et al., 2000). This indicates that stimulus functions can be altered in at least two separate ways: They can change in the way long described within operant and respondent conditioning, and they can also be altered, or transformed, through derived relational responding. In the first case, stimulus functions are established through contingencies: direct relations between stimuli and/or physical similarity between stimuli (generalization). Within RFT, these are usually called *direct contingencies* to distinguish them from derived relations. Stimulus functions that arise through direct contingencies are called *direct stimulus functions,* and those that are established through derived relations between stimuli are called *indirect* or *derived stimulus functions.*

This brings us to the answer to the second question posed at the beginning of this chapter, regarding how verbal behavior interacts with human behavior as a whole. *The effect that derived relational responding has on human behavior as a whole is a result of the fact that when derived stimulus relations are established, stimulus functions are altered.* RFT refers to this as *transformation of stimulus functions.* In ordinary language, this means that something that has had a certain meaning acquires a new one. For example, events that had been neutral take on the function of triggering anxiety, an event that once functioned as a discriminative stimulus for drawing closer now becomes discriminative for avoidance, or a previously neutral event becomes reinforcing or punishing, and all of this can occur *in the absence of direct contingencies* for such new learning.

This means that new and flexible paths for learning have been opened. But it also raises the third question posed at the beginning of this chapter: What governs this behavior? Based on the fundamentals of behavior analysis, it is not enough to know that this is an ability human beings learn and that it has certain consequences. We need to know how contextual factors influence this behavior.

THE ABILITY TO RELATE EVENTS AND STIMULI ARBITRARILY

The behavior of most organisms can be governed by a minor aspect of the context they are in. In an experiment with pigeons, for example, the specific color of a key can govern the pigeon's pecking at it. The pigeon learns to peck a key when it is lit up in red, but not when it is lit up in green. This does not, however, imply that other aspects of the context are completely insignificant. The color by itself does not necessarily exercise control. If something else in the pigeon's environment is red, the pigeon may not peck at it if it is too dissimilar to the key in some other respect. But it would be possible to train the pigeon to peck at anything that is red. The color red would then be a completely decisive contextual factor for controlling what the pigeon pecks at. We could say that the pigeon has learned to abstract this particular contextual factor and acts on or reacts to it. We also know that many animals can learn to respond to abstracted relations: relations based on the physical properties of stimuli. A monkey can, for example, learn to always choose the longest stick (Harmon, Strong, & Pasnak, 1982). If we consistently reinforce a choice of the longest of several sticks, in time the monkey will choose a stick that it has never been rewarded for choosing simply because it is the longest stick. This could be the case even if there is an alternative stick available and the choice of that stick has previously been reinforced because it was formerly the longest in a set of choices. The monkey is thus acting on an abstracted relation between the sticks. We might say it is acting on the relation in itself. Note, however, that the monkey is still acting based on the physical properties of the sticks—specifically, their length. Human beings seem to be able to take this ability of abstracting features of the environment a step further. At an early age, we learn to respond to relations that are not governed by the events that are being related. The events that are related need not be contingent with each other, either in the present or earlier in our history. And they need not be related according to physical properties either. Instead, the relating is governed by other factors in the specific context. We learn early on to abstract other contextual factors or cues that govern how things are related, independent of direct contingencies or physical properties of stimuli. This skill of relating based on specific contextual cues explains how relations can occur indirectly—without any direct learning—as described above in connection with various experiments. This skill has far-reaching implications for our behavior, since the relations that are established in this way transform stimulus functions and thereby give events and stimuli functions that they do not inherently have in themselves. The fact that relations between stimuli can be governed by contextual cues other than the stimuli being related thus explains how derived stimulus relations arise.

Here's an example. Contemplate the two figures # and @. Let's start by reflecting on what direct relations exist between these two figures. The way you, as a reader, see them in print, there is a relation of dissimilarity between them, based on their physical properties. This is true in several ways: For one, # is smaller than @. And since this text is in English, we can assume that readers do their reading from left to right, so the figures have a relation of # preceding @. This is because all readers of English have a common learning history in this respect. Are there any additional direct contingencies between these two that are a direct result of the reader's learning history? To an individual reader, there are a number of different relations that might apply, depending on the person's specific learning history. Either of the figures could, in theory, be related to some kind of uneasiness because of a direct contingency in a specific reader's history, while the other figure could be connected to something pleasant.

I will now establish a new relation involving the figures # and @, independent of the direct contingencies described above. I'll do this by adding a typical contextual cue that all readers will respond to: # is larger than @. That this relation can be established regardless of the physical properties of the stimuli related is evident, in this case by the fact that the new relation is in some ways contradictory, since readers actually perceive @ as the larger figure. Despite this, no reader has any problem in acting or reacting in accordance with the newly established relation. The following shows this: Let's give one of the figures a specific function by saying that @ is a sum of money. If you could now choose between # and @, which one would you choose?

At an early age, humans learn to relate stimuli in a way that is not necessarily governed by stimuli being contiguous (in the present or historically) or by physical properties of the stimuli that are related. This finally brings us to the answer to the third question posed at the beginning of this chapter: What are the factors that govern verbal behavior? *This special way of relating is governed by contextual cues that specify the relation regardless of the properties of the stimuli that are being related.* There is an example of such a cue in the text above: "is larger than." The stimuli that are used in this way are mainly sound combinations in the shape of what we call words and sentences (or visual combinations in the case of written text), but other stimuli, like different gestures, can function in this way as well. Because the contextual cues that govern which relation is established can be independent of the stimuli that are related, the relation becomes arbitrarily applicable. Anything can be put in relation to anything else. Since derived relations are established by arbitrary stimuli that are agreed upon by the social context, in RFT these relations are often called *arbitrary relations*. Correspondingly, relations that are based on contiguity between stimuli or on formal, physical properties of the stimuli that are being related (as is generalization) are called *nonarbitrary relations*.

In everyday life, arbitrary and nonarbitrary relations are constantly combining to affect stimulus functions and thus human behavior. Therefore, in everyday examples it is often impossible to determine whether stimulus functions were established directly or by derived relational responding. However, from experimental work we do know that verbally competent individuals derive relations in the way described, and that such relations occur in much larger numbers than directly trained relations do (Wilson, O'Donohue, & Hayes, 2001). It is therefore reasonable to argue that the majority of the relations that arise in a certain linguistic context are derived relations. At the same time, interaction of the two kinds of relations is an ongoing process. The example involving Kyle showed how the relation "chair" → "stol" was established not by direct training, but through derived relations. However, if Kyle learns to use this Swedish word as a variant of the English word "chair," a direct relation will gradually be established, since he is using the words in connection with each other. This is despite the fact that the relation originated as a derived relational response.

HOW DO WE LEARN TO RELATE EVENTS ARBITRARILY?

We have only limited knowledge about the learning history that results in the behavioral repertoire discussed in this chapter. A great deal of research is still needed in order to elucidate the details. The following account of how we learn to relate events arbitrarily is based partly on the existing general knowledge about how language learning occurs in a natural context, and partly on studies of children going through the specific training that I am about to describe (McHugh, Barnes-Holmes, & Barnes Holmes, 2004; Heagle & Rehfeldt, 2006; Luciano et al., 2007).

I'll start with the relation that has dominated my description thus far: coordination. This is a relation of similarity, equivalence, or "same as." This is the relation between stimuli that we refer to in everyday language as symbolization: One specific thing stands for another. This is the relation described in Murray Sidman's classic experiments, and the relation that is established in the example of Kyle, who learns to relate different new words to "chair." Coordination is the first relation we learn, and it is easy to see that it is absolutely fundamental to human language (S. C. Hayes, Barnes-Holmes, & Roche, 2001). A sound or a series of written letters stands for something, refers to an object or an event, and is in certain respects interchangeable with that "something" referred to. For example, the word "shoe" is in a relation of coordination to an actual shoe.

Early in children's lives, typical contextual cues used to teach them to relate things by way of coordination are words like "is," "same as," "goes together with," and "means." And particularly when it comes to children with language difficulties, the training is most advisably begun using objects that have a formal similarity to each other—stimuli that have a nonarbitrary relation, such as a couple of toys, everyday objects, or pictures (Luciano, Rodriguez, et al., 2009). These are presented to the child while the adult who is training the child simultaneously brings in the contextual factors by saying, "This goes together with this," "This is the same as that," and so on. When the child orients toward one of the objects based on the other object, in line with the relation the adult is trying to establish, this is reinforced: "Good, those are the same," "That's the right one," and so on. The training is done in both directions and using the same contextual cues, in this case sound combinations. It is important that the training be done in both directions—and it is also what typically takes place very early in children's natural language training. At seeing her father, the child hears "Daddy—look, there's Daddy," and after the question "Where is Daddy? Where's Daddy?" the child's behavior is reinforced if she orients toward her father. In this last example, there is no formal similarity between the word and the object (the dad), but the direct contingencies that have been established by ongoing repetition and association are used, and they provide a direct relation through respondent learning. After the child has seen her father a large number of times while simultaneously hearing the word "Daddy," this word—through respondent conditioning—will trigger a perceptual experience in the child similar to actually seeing her father, just like I described earlier in connection with seeing a red car. Skinner referred to this as conditioned seeing, or seeing in the absence of the thing seen (Skinner, 1953, 1974).

In formal language training, a number of different stimuli are used: "Where is the ball? Look, there's the ball." "Where's the kitty? There's the kitty." If the child orients toward these stimuli or does any other action in relation to these stimuli, this behavior is reinforced. The training is performed in both directions (word "ball" → actual ball; actual ball → word "ball") while the contextual cues—sound combinations as well as movements and gestures—remain consistent. Although the objects vary, certain contextual cues stay the same. Gradually, the child abstracts these contextual cues as decisive for the relation between different stimuli. This means the training can move on, using new stimuli that lack physical similarities and also have no previously trained relation in the child's learning history. The same contextual cues ("is," "goes together with," "same as") are still used. If the child acts based on the objects being the same or going together, this is reinforced: "Yes, that's right!" In this way, the child learns to relate stimuli that are new and that lack formal similarities. Once she does this based solely on the arbitrarily established contextual cues the social environment provides, we have an arbitrarily established relation of coordination. Suddenly, based on a certain socially produced context (mainly sound combinations) being present, one thing can mean another, regardless of the individual's earlier history in connection with the stimuli involved or the physical similarities between them. If a four-year-old boy has experienced riding in a small rowboat and has learned to call it a boat but has never experienced going by ferry, the word "ferry" can take on a number of functions for him when he is told that a ferry is a large boat. Once this repertoire is in place, the child will also successively learn to do this silently, according to the principles described in chapter 2.

In a similar way, the social environment trains us in the ability to arbitrarily put events or stimuli into other types of relations, such as comparative relations (more/less). This is the type of relation described above, using # and @. In natural environments, there are plenty of possibilities for letting children encounter nonarbitrary comparative relations between objects that naturally vary in regard to different physical characteristics, and these can be used as opportunities for training. For instance, some things are bigger or smaller than others, heavier or lighter than others, or noisier or quieter than others. Note that comparative relations often include a relation of opposition. For example, if something is heavier than something else, you might say that this implies a form of opposition: One object is heavy as opposed to the other object, which is light. This might indicate that the natural learning sequence involves learning the arbitrary relation of opposition before learning to put stimuli in a comparative relation. This is also why it is recommended that deliberate training be done in this order (Luciano, Valdivia-Salas, Cabello-Luque, & Hernández, 2009). Much research is needed to understand the process whereby learning a certain type of arbitrary relation is predicated on having learned another arbitrary relation, to clarify both what the naturally occurring sequences are and how training can be done most optimally. However, there are some sequences that seem obvious. For example, in order to be able to learn an arbitrary hierarchical relation, it is plausible that the ability to relate stimuli as both similar and dissimilar is required.

As with training in relations of coordination, training in comparative relations requires us to start with stimuli that are in a nonarbitrary relation based on physical properties. For example, if more/less is the relation to be trained, we might use two boxes with different amounts of the same contents. In this case, the contextual cues would be phrases such as "Which one has more?" "Which one has less?" "Pick the one with the most," and "Does this have more or less?" Again, it is important to train in both directions and reinforce responses wherein the child acts in accordance with the nonarbitrary relation. That is, at this point the trainer should reinforce correct responses indicating both that one has more and that one has less. The training is done using different objects where the nonarbitrary relation is the same, in this case maybe also with greater or smaller piles of marbles, and consistently using the same contextual cues. When the child responds correctly—in line with the contextual cues and the nonarbitrary relation between the objects—the training shifts to arbitrary stimuli, that is, to objects that are not, by physical characteristics, in the relation that is being trained. The child may be shown two objects of the same size and then be asked, for example, "If both of these are sweets and this one [indicating one of the same-size objects] is more than that one [indicating the other object], which one would you like?" The exercises are repeated again, allowing the child to act in relation to objects and events based not on their physical properties, but on relations that are in line with the arbitrarily, socially established contextual cues; in this case, the words "more" and "less." When the child responds in accordance with the contextual cues, independent of the nonarbitrary relations between stimuli, this is reinforced. This same type of training takes place naturally when, for example, children learn the relation between coins.

MANY DIFFERENT RELATIONS BETWEEN STIMULI

There are many different types of relations beyond those described thus far (coordination and comparative): difference (distinction), spatial relations (behind/in front of, above/below), temporal relations (before/after), causal relations (if-then), hierarchical relations ("a part of"), and "relations of perspective" (I/you, here/there).

Most of these relations correspond to nonarbitrary relations that are present between stimuli or events in certain situations. Some things are larger than other things; one variety of trees is, for example, typically larger than another type of tree. A tree is not a stone (distinction). A book is placed on top of another book (spatial relation). I enter the room before you (temporal relation), and if I stand up, it is easier for you to see me (causal relation). The second floor is a part of the building (hierarchical relation). All of these are nonarbitrary relations that we may encounter and that variously influence the functions these stimuli have for us—as well as the function they will have for us when we contact them, or similar stimuli, in the future. But people's behavior is not governed simply by these nonarbitrary relations between stimuli. As discussed previously, for us humans, relations between events can be established arbitrarily, by social whim. These relations can, in a given situation, be established in a way that is independent of direct contingencies between stimuli or their properties. Instead, these relations are governed by arbitrary contextual cues established by the social community. Since relations between stimuli govern stimulus functions, these can become available to an individual in a much more flexible way than if nonarbitrary relations were required. In this way, stimulus functions become "movable" by the social community. We can treat certain events as if they are more than other events and as if they occur before or after other events. We can treat them as if they are a part of other events and as if they cause other events. We can do this regardless of the direct contingencies between these different events.

If the relation "X is greater than Y" can be established for an individual regardless of the nonarbitrary relation between X and Y in the specific situation or in the individual's history, this opens up entirely new prospects for humans to mutually influence each other's behavior. For example, if the temporal relation "Q precedes P" is established, an individual will be able to act in relation to Q and P according to this; perhaps she would choose to focus on Q first, even though she has never encountered Q or P in any such relation, or even though she has actually encountered P before Q in a previous context. Having learned this repertoire, possibilities are opened to the human species which, as far as we know, are unavailable to other animals.

Temporal and comparative relations

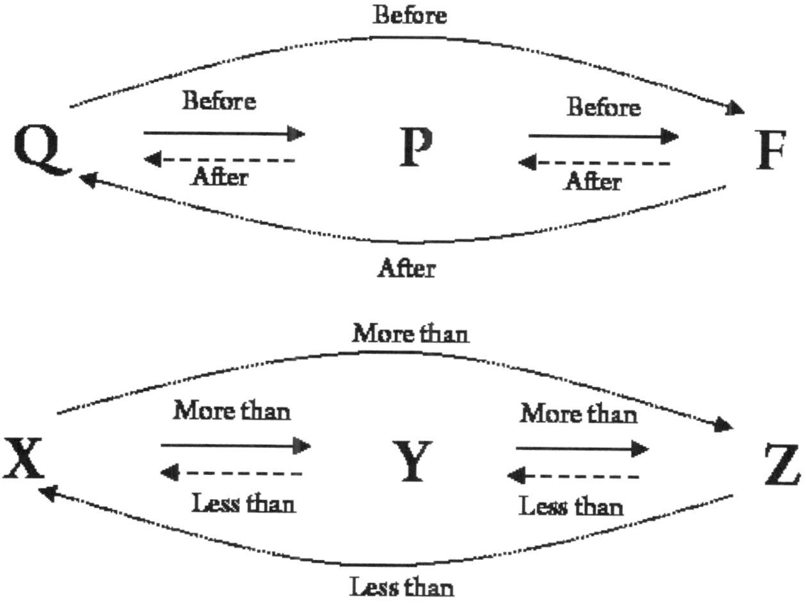

Directly trained ————————

Mutual entailment - - - - - - - -

Combinatorial mutual entailment ....

A number of studies have shown that we are able to put stimuli or events in arbitrary relations other than coordination. These studies are performed by creating experimental analogues of the type of contextual cues that—in a natural environment—govern what kind of relation is established (that is, words like "larger," "in front of," "after," and so on). These analogues are stimuli (cues) that, outside of the experiment, lack the function established in the experiment. For example, a stimulus of a certain shape or color is given the function that the phrase "more than" has in a natural environment. It is then possible to show that these stimuli govern how participants put other arbitrary stimuli in particular relations.

Let's briefly look at one such experiment (Dougher et al., 2007). This is an extension of an experiment I described earlier, involving a mild electric shock, which showed how respondent functions were transformed when stimuli were put in a relation of coordination. One stimulus became equivalent to another by way of a derived relation, and its function was transformed according to this relation. The experiment I am about to describe shows that the stimulus functions can also be transformed according to a relation of comparison.

In the study's first phase, meaningless visual stimuli were established as contextual cues for larger and smaller, respectively. This was done by alternately displaying three different stimuli of this kind to the participants on a computer screen; let's call the stimuli D, E, and F. The participants were instructed to then select one of three comparison stimuli, which were simultaneously displayed in a lower part of the screen. These comparison stimuli were identical in form but varied in size, along these lines: %, %, %, and &, &, &. As the participants chose one of these, they were told whether their choice was "correct" or "wrong." The experiment was designed in a way that made it systematically correct to choose the smallest of the otherwise identical figures if D was displayed at the top of the screen; the largest figure if F was displayed; and the middle-sized figure if E was displayed. This was carried out in a consistent way throughout a series of tasks, with the presentation arranged so that no qualities other than the differing sizes of the stimuli affected the participants' choices.

The experiment had created three contextual cues that could be said to have taken on the meaning D < E < F. Next, several testing phases were carried out, demonstrating that the participants were acting in accordance with the contextual cues displayed on the screen. In one of these experiments, E was given a respondent function by being followed by a mild electrical shock, just as in the earlier experiment (Dougher et al., 1994). Afterward, participants were asked to stay in front of the computer and pay attention to the different figures displayed on the screen. The researchers gauged participants' skin conductance. There was a clear reaction when E was displayed, just as would be expected based on respondent conditioning principles. When F was displayed, the skin conductance was at a higher level than when E was displayed, even though only E had been directly connected to an electrical shock and F had never been directly connected to an electrical shock or to other stimuli that were. In fact, one of the participants pulled off the electrode when F was presented on the screen. The severity of the electrical shock used in the experiment was calibrated to each individual participant to make it slightly unpleasant but not painful. The person in question thus had this experience because of the ability to relate stimuli and, specifically, the aspects of "more than" and "less than."

The researchers came to the conclusion that this once again demonstrated an alteration of stimulus functions based on a derived relation. They also concluded that this alteration meant not only that a stimulus had acquired the function of another stimulus (as in a relation of coordination or equivalence), but that the transformation of stimulus function involved was governed by the specific relation of "more than," which was established in the experiment. Other relations between stimuli have been studied in a corresponding way and have been shown to influence stimulus functions as well. Two examples are opposition (Dymond & Barnes, 1996; Whelan & Barnes-Holmes, 2004a) and perspective (McHugh, Barnes-Holmes, & Barnes-Holmes, 2003).

RELATIONAL FRAMING

Relating in the particular way described in this chapter is termed *relational framing*. A more technical term for this is *arbitrarily applicable relational responding* (AARR). This behavior is a generalized operant. The term "generalized" is often used to describe operant behavior when it is important to emphasize that a particular behavior can only be described in a functional sense, and that it completely lacks topographical description. Other examples of such behavior are "imitation" or "doing something novel." Any type of action could belong to these categories, regardless of its appearance or topography. Relational framing is also this type of action (generalized operant behavior). When we speak of framing things in different types of relations (opposition, comparison, spatial, temporal, and so on), the term "relational frames" is metaphoric. It refers to the way a frame can contain anything. This term does not imply that relational frames exist as mental objects. It is a way of saying that people can put things in various types of relations; that is, we place them inside frames. Obviously, the placing, too, is metaphorical. This relating is not based on any formal or physical properties of the related stimuli; rather, the relations come about as a result of this specific form of human behavior, which in turn is controlled by contextual cues. Relational framing is a behavior that humans learn early in life, through operant conditioning, and is characterized by the three phenomena I have already mentioned:

- Mutual entailment
- Combinatorial mutual entailment
- Transformation of stimulus functions according to the established relation

Combinatorial mutual entailment includes three or more stimuli, and I need to further explain the reason why this is a necessary aspect to include in the definition of relational framing. As mentioned earlier, one aspect of derived relations is that they are mutually entailing, so that if "chair" → "silla," then "silla" → "chair." In this case the entailed, derived relation is identical with the relation that has been specifically trained. If "chair" is the same as "silla," this entails "silla" being the same as "chair." This is how a relation of coordination works, and this also applies to relations of opposition. If Larry is Ben's opposite in some respect, then, in that same respect, Ben is Larry's opposite. Clearly, mutually entailed relations are not always identical: If Larry is bigger than Ben, this obviously entails that Ben is smaller than Larry, not that he is bigger. If I arrive before you do, this entails that you arrive after I do, not that you also arrive in advance.

A further key point is that as long as we are describing mutually entailed relations, both relations have the same level of precision. When relations are combined, however, the precision of the combinatorially entailed relation may be unspecified. If Larry is unlike Ben and Peter is also unlike Ben, we have no way of knowing what relation exists between Larry and Peter. This lack of precision is a given, and the combinatorially entailed relation is therefore defined as unclear.

Combinatorial mutual entailment also illustrates how complex networks of relations are built. If Larry is the opposite of Ben and Ben is the opposite of Peter in that same regard, then Peter is similar to Larry. Then, if we learn that Caleb, Victor, Grant, and Stanley are similar to Ben and that Peter is a disagreeable person, we suddenly know how to relate to all of those involved.

The reinforcing contingencies that we encounter when learning relational framing are of a general and social nature; that is, they are generalized reinforcers, as was described in regard to verbal operants in chapter 2. In addition, relational framing itself gradually establishes a generalized reinforcer that acquires its function precisely through this verbal learning process. A fundamental quality of human languaging is that it must be coherent. We cannot switch words, meanings, or grammatical rules in an arbitrary way. Although the system is arbitrary at its base, once it is established we have to play by the rules. You may be able to break the rules in moderation, but all in all the system's tolerance for this is negligible. The word "boat" cannot first mean one thing, then another, and finally a third. It does not work to first mean one thing by placing "not" in a phrase and then give it a function of "above" in the following sentence. If meaning is to be conveyed, the structure of sentences cannot be changed in an arbitrary way. Coherence is thus built into human language. Provided that languaging (relational framing) is reinforced from an early age, coherence is thereby also reinforced. There are many reasons for the social environment to reinforce coherence specifically as part of language training. As a result, coherence, or the experience of things being logically consistent, becomes a generalized reinforcer for verbally competent human beings (Blackledge, Moran, & Ellis, 2008; Healy et al., 2000).

TWO ASPECTS OF CONTEXT THAT GOVERN RELATIONAL FRAMING

In many of the experiments described in this chapter, the researchers used stimuli that were meaningless to the study participants at the outset. The stimuli had very few stimulus functions, and those they had were restricted. When some new functions, for example discriminative or respondent functions, are trained, they dominate completely. This makes it easy to see that these are the functions that are established for other stimuli in the experiment through relational framing. In real life the situation is much more complex. Most stimuli have a number of different potential functions, based on their physical properties, the direct relations established in the individual's history, and the individual's earlier history of relational framing. When stimuli are put in relation via the type of contextual cues mentioned above ("[...] is [...]," "[...] is located behind [...]," "[...] comes after [...]," and so on), the specific stimulus functions that come into play are selected from a wide array of possible functions. To say this more technically, given a particular relation, only certain stimulus functions among the many possibilities are transformed.

Here's an example: If we were to talk about an apple, I think that each and every reader would find that the word "apple" is in a relation of coordination with a real apple. This specific type of relation (words stand for what they refer to) is so solidly learned that all that is needed to establish this relation is the context of talking to one another. You could put it like this: We presuppose that people mean something by what they say. But we can still vary which of the many possible stimulus functions of apples come into play in our dialogue without altering the fundamental relation of coordination between the word "apple" and a real apple in the reader's history. Read the following sentence, then stop for a moment and make a note of what you experience:

Imagine an apple.

Now, compare that experience with what you notice while reading this sentence:

Imagine eating an apple.

In both examples, the context (in this case the rest of the sentence) established around the word "apple" cues the same kind of relation between the word "apple" and the reader's experience of a real apple: a relation of coordination. But the functions that are brought to bear for the reader probably aren't identical for the two sentences. Your imaginary contact with an apple shifted, meaning that different stimulus functions were contacted, even though the same relation was cued. The socially and arbitrarily created context not only cues a particular relation, but also determines which stimulus functions, out of many potential functions, are transformed by this relation. The words "imagine" in the first sentence and "imagine eating" in the second cued different transformations of functions. The change in the context that the reader encountered (in this case, a sentence) transformed which functions of a real-life apple were brought to the fore for the reader. Other functions of an apple would be contacted with yet another contextual change:

Imagine sitting on top of an apple.

Relational framing is always dependent on both of these aspects. On one hand, a certain relation is established between the stimuli, and on the other, certain stimulus functions are selected in accordance with the established relation. Both aspects are governed by contextual cues and can therefore be established arbitrarily. In the above example, a number of factors—factors that are more or less always present when we read a text in our own language—controlled which relation was established between a real apple and the word "apple." Words in print are in a relation of coordination with the things they refer to. Out of the many possible functions that were made available based on this relation of coordination, different functions were selected as a result of the differences between the sentences. In RFT, a contextual cue that governs which relation is established between stimuli at a given moment is designated C_{rel} (from context and relation). A cue that governs which function is selected based on this relation is designated C_{func} (from context and function). Here's another example: "Larry is better than Peter at playing tennis." The words "better than" most likely function as C_{rel} here, as they control which relation is established between Larry and Peter. The words "at playing tennis" function as C_{func} by governing which function is brought to bear, in this case through a comparative relation.

C_{rel} and C_{func} are functional classes, so we cannot say that any given factor can always be designated as one or the other. To make this determination, we have to analyze a given process or event, as is always the case for a functional analysis. The distinction between contextual cues that govern the relation and contextual cues that govern which stimulus functions among the many possibilities are transformed is not always an essential one, but it can be. One situation where the distinction is important is in certain clinical interventions. This is something I'll return to in part 3 of the book.

A NEW DEFINITION OF VERBAL BEHAVIOR

I have now answered the three questions with which I opened this chapter—not altogether exhaustively, but hopefully to the extent needed to provide a basic understanding of key concepts in relational frame theory. The description of derived relational responding and the theory that has developed based on this phenomenon provide a new definition of verbal behavior—a definition that is more precise than what Skinner offered (1957). To frame events or stimuli relationally, according to the three criteria mentioned (mutual entailment, combinatorial mutual entailment, and transformation of stimulus functions) is the behavior that RFT refers to as verbal (S. C. Hayes, Fox, et al., 2001). In a verbal interaction, this is done by both the speaker and the listener. We can also do this silently to ourselves, in which case the individual becomes both speaker and listener. Verbal stimuli, according to RFT, are stimuli that have their effects because they participate in relational frames (S. C. Hayes, Fox, et al., 2001).

The data that this definition is based on were not available when Skinner wrote his texts about verbal behavior. However, the phenomena he described are still important even though, according to RFT, they are to be understood from this new perspective. Skinner described the difference between behavior that is governed by direct contingencies and behavior that is governed in other ways. For the latter he used the concept of "rule-governed behavior," but he never explained how such behavior takes shape. As discussed in chapter 2, there are some gaps in his analysis, and leading behavior analysts have pointed this out. These gaps are filled by RFT by way of its new definition of verbal behavior, which makes it possible to define rule-governed behavior precisely. (I will return to this in chapter 6.) It is important to note that, under this new definition, a specific behavior that is verbal according to Skinner's definition may or may not be verbal according to RFT. For example, a tact as defined by Skinner is not necessarily verbal from the perspective of RFT. If a child tacts "a dog," this response could be the result of having seen a dog at an earlier point and having uttered "a dog" and then receiving reinforcement that was contingent upon this response. Yet all of this can occur without the phrase "a dog" participating in a relational frame, in which case the tact is not verbal according to RFT's definition because it has been established solely through direct contingencies. Many times, though, when children use the phrase "a dog," it is also in derived relations with real dogs (and other things), and the child's response is then verbal in the RFT sense, as well. A similar distinction between the Skinnerian definition of verbal behavior and the RFT definition can be made for the other verbal operants described by Skinner. (For a more detailed comparison of Skinner's definitions and those of RFT, see D. Barnes-Holmes, Barnes-Holmes, & Cullinan, 2000.)

SUMMARY

Verbal behavior, according to RFT, is to put stimuli (events) in relation and to act on or react to stimuli based on the resulting relations. This behavior is learned in early language training and is a generalized operant. This particular way of relating is primarily controlled not by the nonarbitrary relation between the stimuli being related, but by other contextual cues. The stimuli that function as such cues are mostly sound combinations—what we would normally call words. But other contextual factors, such as gestures or even features of the nonsocial environment, can also have this function. This means that anything can be related to anything else, since the social environment can manipulate the governing contextual cues.

This behavior is called relational framing, or, more technically, arbitrarily applicable relational responding (AARR). The latter term illustrates several of the considerations discussed in this chapter. "Responding" makes clear that this is a behavior. "Relational responding" lets us know that this behavior involves relating events to each other. That these relational responses are arbitrarily applicable tells us that this relational responding is not based on any nonarbitrary or formal (physical) relations between the stimuli being related; rather, it is based on aspects of the context that specify the relation such that the relational response can be brought to bear on any stimuli or events, regardless of their nonarbitrary properties (Stewart & McElwee, 2009). These aspects of the context, or contextual cues, are established by social whim (hence arbitrarily applicable). Thus anything can in principle be put into any frame. "Arbitrarily applicable relational responding" and "relational framing" are synonymous. An additional synonymous term is the one used in this chapter's title: "derived relational responding."

For humans, this behavior changes the possibilities for learning in a crucial way, allowing us to manipulate stimulus functions in a way that would not be possible if stimulus functions were solely dependent on nonarbitrary relations or contingencies between stimuli. Once derived relational responding is learned and exhibited, stimulus functions can be changed in a moment through these responses. What was punishing can become reinforcing, and vice versa. Something we have never encountered before can function like a discriminative stimulus for a certain behavior. An event can suddenly acquire respondent-type functions that it has never had before and thus, for example, change one's emotional state. And all of this is based on socially invented cues, which, on one hand, govern the relations that are established and, on the other, govern which specific functions are selected based on these relations.

Before I turn to some of the complex behaviors that this foundation makes possible, I would like to go back to the topic of this book: the power of thinking. From the viewpoint of radical behaviorism and RFT, what we usually call thinking is not a behavior in a class by itself. Thinking is not seen as something that takes place in a mental world. Neither is it seen as something that follows principles other than those of any outwardly observable behavior. As I described in chapter 2, thinking is simply verbal behavior performed in such a way that the only possible observer is the person who is performing it—the person doing the thinking. That definition stands. The power of thinking, then, is in the power of relational framing. Based on what we first learn from our social environment according to the principles described in chapter 2, it is possible for the individual to gradually perform this same behavior silently to herself. The increased flexibility that is added to human behavior in general by arbitrarily applicable relational responding also allows an increased flexibility in private behavior. The power of thinking lies precisely in this ability to relate anything to just about anything else.

CHAPTER 5

Analogies, Metaphors, and Our Experience of Self

Chapter 4 outlined the fundamental principles of arbitrarily applicable relational responding. The essentials are actually rather straightforward: We can relate events by social whim, making it possible to change stimulus functions in a very flexible way and thereby change our own and other people's behavior. Our languaging can be likened to a social game we play together. Most people learn the basic aspects and conventions of this game at an early age. If it were a board game, we could say our pieces (different languages) vary a bit in color, but the gist of the game is the same. It is built on one fundamental principle: that relating events in an arbitrarily applicable way is itself a learned behavior. The principle is straightforward, but its consequences are dramatic for us humans. Once the basic skill is learned, an exponential increase in complexity opens up for us. Once we are capable of relating words and other events at social whim (arbitrarily) we can bring together events or experiences in a totally new way to influence behavior. For example, the word "language" is in coordination with our direct experience of languaging, and the word "game" is in coordination with our direct experience of playing games; so, just by relating these words, we can put two different experiences in relation to each other and influence further behavior, perhaps by saying, "Language is a game we play." This is a specific type of analogy: a metaphor. In everyday language, we would say that this process gives something a new meaning. In the scientific language used in this book, we would say that by relating events, we are transforming their stimulus function.

Many researchers and theorists at the intersection of psychology, philosophy, and linguistics believe that the ability to create analogies and metaphors is at the root of human language and that it forms the very foundation of our way of speaking with each other (Barclay, 1997). A leading researcher put it this way: "All knowledge is ultimately rooted in metaphorical (or analogical) modes of perception and thought" (Leary, 1990, p. 2). This cross-disciplinary field is vast and embraces many theories, as well as conflicting points of view. My goal here is to establish that there is a scientific consensus of opinion that analogies and metaphors are significant in understanding human language and cognition, and that RFT both contributes to an understanding of these phenomena and shows us how they arise and are influenced.

The observant reader may have noted that many of the relations and functions in the example "Language is a game we play" are not arbitrary or established by social whim. Playing games is probably an actual experience to the reader, just as the experience of languaging is. And there is a relation between these experiences that is not arbitrarily established: a type of similarity. That is what the metaphor is making use of, what it is pointing to. This is an important observation. The fact that we have the ability to relate events arbitrarily, demonstrated in the laboratory experiments described earlier, does not mean that we relate events in a completely arbitrary way most of the time. We have the repertoire to do so, and this occurs in its most pure form in certain types of mathematics and abstract logic. But in most situations we use arbitrarily applicable relations in combination with relations established through direct contingencies and formal properties of stimuli. This interplay between direct and derived relations is what gives verbal behavior its impact. Or, to express this more technically, arbitrarily applicable relational responding transforms stimulus functions, many of which were originally established through direct contingencies. This explains the enormous impact of verbal behavior. Imagine little John, who has been frightened by a mean dog. The concept of respondent conditioning makes it easy to understand his fear of similar dogs. But if John refuses to visit Aunt Patty's house (where he has never seen a dog, much less been frightened by one) because someone said, "Her cat is like an old dog," this is because relational framing has brought him into contact with some of the stimulus functions of his earlier direct experience with a mean dog.

If relational framing is the fundamental process in human language, it follows that it influences most spheres of human experience, since language is basically omnipresent. If the claims of RFT are correct, the potential applications are almost infinite. To give just a few examples, relational framing should be relevant to fields as diverse as education, psychological development, the training of children with specific language difficulties, linguistic research, politics, social processes, existential issues, and individual psychological problems (S. C. Hayes, Barnes-Holmes, & Roche, 2001; Rehfeldt & Barnes-Holmes, 2009). However, this book restricts itself to applications in the field we normally call psychotherapy.

Before I can go on to the applications related to psychological problems and psychological therapy, I need to describe a few more key concepts in RFT. The rest of this chapter is dedicated to two of these fundaments: first, as already touched upon, the fact that arbitrarily applicable relational responding can be used to relate relations to other relations and thereby create what we call analogies and metaphors; and second, how this behavior creates our experience of self.

ANALOGIES

As mentioned above, the ability to think analogically is widely seen as being of fundamental importance to human language and cognition. In an anthology on analogical reasoning written by researchers within different fields of application (Vosniadou & Ortony, 1989), several contributors discuss how we are to understand the different types of comparison and "mapping" (a common term in cognitive models of analogical thinking) that are constituent parts of analogies (Collins & Burstein, 1989; Gentner, 1989). The core idea is that knowledge is transferred from one field of experience (often called "base" or "vehicle") to another (usually called "target"). The base is the field of experience which is most familiar; the target is then the field where knowledge is to be expanded, which is done by relating the two fields. The well-known analogy between the solar system and an atom is a classical example. When using this analogy, we assume that the solar system is more familiar, so it functions as the base, with the atom being the target. This ability to act on or react to patterns and similarities is thought to be very fundamental to human cognitive abilities and as such is currently the subject of many studies within the wide field of research known as cognitive science. Researchers are trying to design computer models of the phenomenon, searching for its biological bases, and studying how people behave through psychological experiments (Gentner, Holyoak, & Kokinov, 2001). The basic phenomenon is often described as the ability to relate both objects and the relations between objects (Holyoak & Thagard, 1997; Gentner, Bowdle, Wolff, & Boronat, 2001). One problem that arises in trying to familiarize oneself with this field of research is the lack of agreed-upon basic models. It may be an exaggeration to say that each writer or researcher has his own model, but even so, it would be an exaggeration with a point. Many different models are presented, and most seem to lack common points of departure (Collins & Burstein, 1989; Kokinov & Petrov, 2001).

RFT allows us to describe the human behavior that underlies our ability to create analogies. We can do this using thoroughly operationalized concepts in the way that has always been the distinguishing feature of behavior analysis.

Analogies originate in relational framing—specifically, the relational framing of relations. Relational networks that are already established, and which in themselves usually consist of both arbitrary and nonarbitrary relations, are related. This is not, in principle, anything novel as compared to what I described in chapter 4. If different stimuli or events can be related arbitrarily, then relations can also be related arbitrarily, following the same principles.

Let's look at a rather uncomplicated and typical analogy: Volvo is to Saab as nectarines are to peaches. Here we have a relation that is already established, the relation we traditionally call the base. Nectarines are in a known relation to peaches. This is a relation of similarity: Both are edible, taste sweet, have a similar shape, grow on trees, and so on. By putting this relation in coordination with the relation between Volvo and Saab—the target of this analogy—stimulus functions are transferred from one relational network to the other. Someone who had just been asked to choose either Volvo or Saab and who thought the difference was quite large now has a new basis for his choice. Through a relation of coordination between two different relations (the relation between nectarines and peaches is in coordination with the relation between Volvo and Saab), the stimulus functions are transformed, which could have an impact on the listener's behavior.

Note that both arbitrary and nonarbitrary relations are a part of the analogy. That the words used refer to certain actual phenomena is arbitrarily established. Certain relations in the respective networks are arbitrarily established as well, like the fact that both nectarines and peaches belong to the category we have learned to call fruits. Likewise, Volvo and Saab are in arbitrarily established relations; for example, both are trademarks of motor vehicles. At the same time, the analogy makes use of nonarbitrary relations of similarity. Nectarines and peaches are similar in certain respects, and Volvo and Saab are also similar in certain respects; both names refer to a car with certain qualities. The analogy is based on an abstraction of a similarity between similarities.

What is fundamental in an analogy, according to RFT, is that a relation of coordination is established between two networks of relations. The relations within these networks are, however, not necessarily relations of coordination. In the example of an atom and the solar system, the analogy points to the spatial relations in the respective networks: One way of envisioning an atom is that certain parts of the atom orbit around other parts, just as in the solar system. This is an abstraction of a similarity between spatial relations in two networks of relations.

Let me give you another example: Susan and Larry are like cats and dogs. Again, two relational networks are put in a relation of coordination. This occurs through the contextual cue "are like." This contextual cue is normally used to establish a relation of coordination between events. In this example, however, the relation the analogy points to within the respective networks is considerably more complicated than pure coordination. You could perhaps say that it is a relation of opposition, but this, too, is an oversimplification. The analogy points to a relation that consists of a number of different types of interactions that we might summarize as "always fighting," "animosity," and the like. In this instance, the relation between cats and dogs is the base, and the relation between Susan and Larry is the target.

This is an example of what we normally call a metaphor or possibly a simile. Most of the time, there is no clear distinction between the concepts of analogy and metaphor, not even in scientific settings. Maybe it is a sign of the usefulness of RFT that the definitions it provides for these two concepts allow us to describe a key difference between analogies and metaphors. I will return to this in the next section.

Let me finish this section on analogies with a short commentary on the existing experimental support for an RFT analysis of this phenomenon of relations between networks of relations, which is highly essential to human language and cognition. For more details, interested readers are referred to the references cited.

There are several studies that illustrate the process described above. In design, they are similar to the design of the experiments on derived relational responding described in chapter 4. Again, they make use of arbitrary stimuli, such as nonsense syllables or abstract shapes, and subjects are trained and tested for development of derived relational networks, typically coordination or equivalence and difference. The experiments then test that subjects can derive relations between the previously established networks of relations. This is done by showing that participants can relate pairs of stimuli to other pairs of stimuli when the relation between the pairs is the same (for example, they are both equivalence relations or they are both difference relations). Using a more recently developed methodology, referred to as the relational evaluation procedure (Stewart, Barnes-Holmes, & Roche, 2004), participants assess the coherence of sets of analogical relational networks by using cues previously established to mean true or false. This ultimately allows for much faster generation of derived analogies than the traditional matching to sample protocol.

These studies can potentially be used both to develop an understanding of how analogies work when they impact us and to show how certain contextual factors govern this behavior. The studies performed thus far have confirmed that analogies function as RFT assumes they do; that is, through a relating of relations between stimuli (Barnes, Hegarty, & Smeets, 1997; D. Barnes-Holmes & Stewart, 2004; Lipkens & Hayes, 2009; Stewart et al., 2004; Stewart, Barnes-Holmes, Roche, & Smeets, 2001, 2002). Several studies also show that, in young people, the ability to create analogies develops in parallel with the overall abilities of relational framing (Carpentier, Smeets, & Barnes-Holmes, 2002; Carpentier, Smeets, Barnes-Holmes, & Stewart, 2004).

METAPHORS

Metaphors are a particular type of analogy, so they also rely on different relational networks being related mutually in coordination. Here's an example of a typical metaphor: "To argue with him is to be run over by a steamroller." This example fits well with the above description of analogies. We have described two events that make up two relational networks between different stimuli. One consists of a person who argues with others, and the other consists of a person being run over by a steamroller. Both networks contain arbitrary relations (such as the relation between the words that are used and the objects they refer to or how it feels to be run over by a steamroller, which most listeners have never experienced and can contact only through verbal functions) and nonarbitrary relations (such as the listener's experience of arguing with others or the direct experience of having seen something being run over by a steamroller). These two are placed in an arbitrary relation of coordination by a contextual cue: "is." Just as with the analogies discussed above, this relation of coordination is nevertheless not exclusively arbitrary. It employs a nonarbitrary similarity between the two networks in at least one particular sense: how it would be to try to stop or influence this person or a steamroller. The two relational networks contain different types of relations, but the relation that exists in both and is related here is a causal relation—that of standing in the way of someone or something on one hand, and the consequences of doing this on the other.

How, then, do we distinguish between a metaphor and an analogy? In the examples of analogies given earlier, there is symmetry between the nonarbitrary relations used in the analogy. In the analogy "An atom is like the solar system," the atom is the target and the solar system is the base, but the analogy could work in the other direction. The nonarbitrary spatial relation applied in the analogy (one part orbits another) is the same for both of the relational networks. If an individual understood an atom in this way (in the simplified sense illustrated by the analogy) but did not know how the solar system works, we could say, "The solar system is like an atom," using the atom as base and the solar system as target. The same thing is true of the example of different cars and different fruits, even though it seems unnatural because it's hard to imagine a situation where someone is unfamiliar with the relation between nectarines and peaches and therefore finds an illustration referring to a known relation between cars helpful. Nevertheless, the two relational networks are also symmetrical in this example; that is, Volvo and Saab are roughly as similar to each other as nectarines and peaches are. The example "Susan and Larry are like cats and dogs" is a borderline case. If someone didn't know in what way cats and dogs relate to each other but had experienced the way in which Susan and Larry fight, the analogy could, at least tolerably, function in the opposite direction, with Susan and Larry as base, and cats and dogs as target.

Compare this to the metaphor "To argue with him is to be run over by a steamroller." The nonarbitrary relation that is applied in this metaphor—the similarity between two events involving standing in the way of an irresistible force—is not a symmetrical relation. One of the two relational networks being related has more of the key quality than the other network does. The base (being run over by a steamroller) has this quality to a greater extent than does the target (arguing with a certain person). True analogies can be used in both directions, as shown in the examples above. This is not the case with metaphors. If we were to turn the described metaphor around and say, "To be run over by a steamroller is to argue with him," this metaphor probably won't function well. The base and the target cannot switch places. In more technical language, certain stimulus functions of what we might call "the futility of resisting" are more pronounced in one of the networks. This is, of course, the one containing the steamroller. This means the stimulus functions of the separate networks are mutually in a nonarbitrary, hierarchical relation. In plain words, they are similar, but the characteristic that is similar is considerably more pronounced in one of the networks.

The preceding discussion may seem like an unnecessarily detailed survey. But this is an essential issue, since this quality of the metaphor, according to RFT, is the very core of its usefulness. Metaphors function by swiftly transferring a characteristic that is highly evident in one event to a different event. Remember that, within RFT, an arbitrary transformation of functions of one stimulus (event) by another is at the very core of human languaging. Metaphors are simply complex networks of relations that achieve this transformation in an efficient way. The nonarbitrary similarity between the two networks in a metaphor may support transformation of functions to an extent that analogical similarity alone might not.

When someone tells me that "sayyara" is the same thing as "car," the stimulus functions of a real car (through the English word "car," which we assume has already acquired some of these stimulus functions) are transferred to the word "sayyara" (an Arabic word for "car," transcribed in Latin characters). When I hear someone say, "Trying to solve this problem by thinking about it more is like trying to plow a field by tilling the soil in your head," this statement has a fundamental similarity to the first example. In each case, two phenomena are related arbitrarily through a contextual cue ("is the same as" and "is like"). However, there is also an important difference between the statements. In the first case ("sayyara" and "car"), two individual stimuli are related, whereas in the second case the relation is between two complex relational networks, each having certain qualities. Each of the two networks includes a certain relation that is the same in both networks. Likewise, both the atom and the solar system contain the spatial relation "one part orbits around another part"; and both Volvo and Saab, as well as nectarines and peaches, have certain similarities. These relations are not arbitrary. It is these nonarbitrary relations, internal to each network, that are related to create an analogy.

A metaphor is an analogy where the qualities in focus are more pronounced in the base than they are in the target. In a good metaphor, the characteristic is typically very obvious in its base and only scarcely noticeable in the target; this subtle characteristic is highlighted precisely through the metaphor. For example, thinking about a problem is something most of us see as a naturally good thing. We are accustomed to solving problems and difficulties in that way. This can lead us to doing this type of thinking in a compulsive way, even in situations where it doesn't actually work. In a situation like this, we may have a faint feeling that this thinking isn't getting us any closer to the solution, and yet it seems natural, like what must be done. When we hear someone say, "Trying to solve this problem by thinking about it more is like trying to plow a field by tilling the soil in your head," an experience is transferred—something we often call an insight—from one relational network to the other. The nonarbitrary relation between these two networks may influence our subsequent actions if it leads to the view that continued thinking about the problem isn't as necessary or effective as we believed.

The preceding analysis of metaphors based on RFT is not merely an intellectual construction or a conceptual model. It has evolved gradually through experimental studies parallel with studies of analogies (Stewart & Barnes-Holmes, 2001, 2008).

Metaphors Are Everywhere

Thus far, the examples of metaphors I've given are easy to follow, and it is easy to identify them as metaphors. However, the use of metaphors has a much wider scope. Metaphoric speech is not limited to clear examples like these. Metaphors are everywhere; they are even woven into ways of speaking we do not immediately identify as metaphoric. I see an example in the previous sentence: woven into. This is something you do when you are weaving. But what I was writing about was how metaphors are inserted everywhere in our way of speaking or writing; I wasn't writing about weaving. "Metaphors are related to our use of language in the same way that yarn is related to a piece of woven fabric" is probably what the reader catches (did you notice that?), and this is the relating of relations to each other described above. The metaphor was embedded (did you notice that?) without my paying (did you notice that?) attention to the metaphor when I wrote it. It may be possible to replace this metaphor—that metaphors are woven into language—with something nonmetaphorical, like the word "insert": Metaphors are inserted in our language even when they do not stand out (did you notice that?) to us. Saying that they are inserted in language may seem like a drier (did you notice that?) way of putting it (did you notice that?) compared to saying that they are woven into language. And we might ponder whether "inserted" is not also metaphorical, if we take a closer look. These types of metaphors, which are an established part of our vocabulary, are usually called frozen metaphors. Of course, the usage of the word "frozen" is yet another example of the phenomenon I have tried to illustrate.

Metaphors are built into the foundation of language. There is scientific consensus about this, and RFT shows us how this foundation is built. It seems like a fitting last example to point to (!) the fact that this section's heading—that metaphors are *everywhere*—is metaphorical too. As concepts, metaphors don't exist anywhere in the physical world; they are simply related to our languaging in the same way that something tangible can be everywhere around us.

Metaphors and the Two Types of Contextual Cues

In chapter 4 I described how the contextual cues that govern relational framing can be divided into two functional classes: those that govern the relation that is derived (C_{rel}) and those that govern which specific functions will be transformed in accordance with that relation (C_{func}). I mention this once again because metaphors are a good example of when the difference between C_{rel} and C_{func} matters.

Let's return to the metaphor "To argue with him is to be run over by a steamroller." There are several elements in the context where this is uttered that function as C_{rel}, most obviously "is." This governs what kind of relation is established between the two events: one of similarity. The real-life similarity between arguing and being run over could also function as C_{rel}. The probability of this is greater if the person who hears the metaphor has some experience of arguing with the person in question, meaning he has been in touch with this similarity independent of the metaphor. The person who hears the metaphor without having any previous experience of the person in question, however, will get a clear idea of the similarity only after having heard the metaphor, which means this factor is less obvious as C_{rel}. The more general similarity between being run over and being in an argument could possibly come into play, but it probably has less significance.

However, something must govern which functions in the metaphor's target are going to be influenced. A relation of coordination is established between the two networks ("is"), but not all of the stimulus functions of the base are transferred to the target. Very few people who hear this metaphor would come to the conclusion that the person in question has a steering wheel, rolls forward, or weighs tons. There is something in the context that governs which functions are transformed, that is, something that plays the role of C_{func}. Here, too, the nonarbitrary similarity (or more precisely, the hierarchical relation) between arguing with someone who is very obstinate, on one hand, and being run over, on the other, comes into play. Which functions are brought to the fore can also be governed by other elements in the described situation, such as what the person said. If the person who hears the metaphor has first witnessed a dispute with the person described by the metaphor, factors in that event will no doubt also serve as C_{func}.

Again, it is not essential to accurately map out exactly what is C_{rel} and what is C_{func}. Though you could possibly do this in a restricted experimental environment, it's hardly possible "out there, in real life." Chances are exceedingly slim that we can map out in such a detailed way the events and processes that we experience or that people tell us about. Our knowledge is not sufficient, and in most cases this type of analysis becomes very speculative. What is important is to understand the principal difference between these two types of contextual control. Remember behavior analysis's fundamental aims: prediction and influence. When it comes to influence, the distinction between C_{rel} and C_{func} is important, because it can be easier to influence a process by altering the contextual cues that govern function rather than by changing those that govern relation. This is relevant to clinical interventions, and I will return to it in part 3 of this book.

BEHAVIOR ANALYSIS AND THE CONCEPT OF SELF

The word "self" has a well-established position in both ordinary and scientific language. We talk about self-image, self-control, self-esteem, and self-confidence. I say and think things about myself. The concept is closely tied to our experience of being a person, an "I." It is a core word in the world of psychotherapy, and one important school within psychoanalysis has the word in the name of its approach, namely self psychology (Kohut, 1971, 1977). However, the word "self" has varying meanings, and it has a commonsense background rather than a scientific one. Within behavior analysis, concepts with such weak definitions are often criticized as unscientific. At the same time, the word "self" often occurs within the different applications of behavior therapy, such as in the concepts of self-control and self-discrimination. Skinner wrote about how this word has often been used in a unscientific way, but also said that the word points to an important part of human experience. In fact, he devotes entire chapters to the concept in some of his most important books (Skinner, 1953, 1974, 1989). Because this experience seems to be universal among humans, we have reason to analyze it if human behavior is what we want to understand. Talking about "self" is, no doubt, clearly a human behavior and Skinner was interested in discovering the background to this phenomenon.

The following simple example can be used as a starting point of an analysis of this phenomenon (Skinner, 1953, p. 265): To say "I see a rainbow" is not the same thing as saying "There is a rainbow." In the first case the individual has identified his own seeing; identifying a rainbow is ancillary. To make use of the terminology outlined in chapter 2, he is tacting his own behavior. Skinner viewed this as the foundation of the concept of "self." We learn very early in life, from our social environment, to talk about our own behavior because this is useful to those around us; in time it also becomes useful to the individual, as described in chapter 2. Parts of this behavior belong in the category Skinner referred to as private, that is, behaviors that can be observed only by the person doing them.

When a young child learns to speak, it is initially through echoing (see chapter 2). Initially, the child uses isolated words when interacting with different objects, but as time passes units emerge, such as "Me get dolly," "Me get ball," "Me happy," "I want...," and so on. Certain parts of these units are shifting and others are permanent. One thing that is permanent is the link between the child's own behavior and the words "I" and "me." The way the child acts in coherence with this, like actually saying "I" or his own name in connection with his own actions, is continuously reinforced by the people in his social environment. Along with this successive use of words that are related to the child's own actions, stimuli that are available only to the child himself will also be correlated both with the word "I" and with other words for himself; words like names, nicknames, and the like. So talking about "I" or "me" starts out as a part of bigger units of speech that are learned through operant conditioning. Later, the smaller unit "I" is selected and coordinated with private events—stimuli detectable only by the child himself (Kohlenberg & Tsai, 1995). In this way, the child learns to tact his own behavior, including private events. Parts of the stimuli present for the child as this happens control the tacting of "I." But according to RFT, something further is needed for the complex phenomenon of self to emerge.

THE EXPERIENCE OF SELF: A RESULT OF PERSPECTIVE TAKING

RFT builds on the above analysis (S. C. Hayes, 1984; S. C. Hayes & Gregg, 2000; S. C. Hayes, Barnes-Holmes, & Roche, 2001) and takes its point of departure in derived relational responding, as described in chapter 4 and in the type of relational framing that establishes a verbal perspective. We utilize different relational frames and fit stimuli into these. We fit events into comparative frames: "Love is greater than hate." We fit them into temporal frames: "Larry will be there first." In the same way, we learn to relate events from a perspective: "What if he sees..." According to RFT, this ability to take a perspective through relational framing is what makes the human experience of self possible.

In line with what I described just above, a child can learn to tact "I" as a part of tacting his own behavior, including private events. However, the complexity of this behavior is considerably increased by relational framing. Just as a child is reinforced for using "I," he will also be reinforced for using the word "here." But the child is also reinforced for using words like "there" and "you." And "there" is always somewhere other than "here," and "here" is always from the point of the view of the child. From this kind of training, it seems that a sense of a unique perspective emerges, which we learn to distinguish from the perspective of others. When we humans experience our own unique perspective (I), we do so from our experience that an alternative perspective (you—or he, she, or it) is possible (Buber, 1970). The same dynamic applies to the two other relations thought to be a part of perspective taking: here/there and now/then. "Here" is dependent on "there" for us to experience it the way we do, and "now" is similarly dependent on "then." If the concepts of "you," "there," and "then" were to vanish (which is hardly even conceivable to any of us), then "I," "here," and "now" would lose their respective qualities. What we now experience or mean by these expressions would cease to exist. The way I experience having a perspective of my own, seeing things from precisely where I am situated, is supported by the experience that a different perspective, seeing from where someone else is situated, is possible.

This discussion is an attempt to describe parts of our experience that are so natural and deeply ingrained that they are difficult to talk and write about. According to RFT, this experience of perspective evolves from three different and yet interdependent relations that we can describe as I/you, here/there, and now/then. The two first are spatial relations, and the third is temporal. These three relations are learned as a part of our ordinary language training through an innumerable variety of everyday questions such as these:

- What are you doing now?
- What am I doing now?
- Where are you going to be then?
- Where am I going to be then?
- What have you got there?
- What do I have here?
- Where are you now?
- Where were you then?

In this type of learning, which includes both questions and answers, much of the context will shift from one occasion to the next: what someone does, what someone has, who has or does something, where someone is, when something was done, and so on. All of this is constantly changing. What remains invariable for any particular person is the perspective itself.

An individual always answers from the perspective of I-here-now. This is the only perspective the individual experiences in a direct way. If you as a reader note your own experience right now, you do this from the perspective of I-here-now. Have you ever experienced anything—seen, heard, or done anything—from a perspective other than I-here-now? Hardly, and yet we act and react as if there were another perspective. We can never experience this other perspective directly. The way we, as individuals, relate to a perspective other than our own has been learned. It is clear that young children have not yet learned this and therefore make obvious mistakes if we ask them for such a perspective. If we ask a young child what someone else at the table is eating, we might get an answer that tells us what the child himself is eating. If we ask what someone who is sitting across the table from the child can see, the child might not answer the question correctly. The ability to see perspectives other than one's own is verbally established; that is, it is learned through arbitrarily applicable relational responding. The types of questions listed above will be asked, and answered, on countless occasions. As the child learns to abstract the different perspectives, these perspectives will then become arbitrarily applicable through the contextual cues that different words become. Words like "there," "you," "he," "she," "they," "then," and "yesterday" will cue derived stimulus functions that are brought in from perspectives other than that of the child himself. Nothing is literally brought in, of course. All of this emerges as a result of the transformation of functions. As this happens, we appear to contact a perspective other than our own due to having learned to fit experiences into relations: I/you, here/there, and now/then. This contact is illusory—we can never take another person's perspective in the way that we take our own. At best, we can imagine another person's perspective. We can also imagine ourselves as if we had another perspective: I-there-then. But wherever we go, no individual can ever actually get to you, there, or then.

A verbally competent person is always in the position of I, here, and now. If I see your perspective, I do so from my own. I can try to imagine what I would have done or what I would do if I were in your situation. But if I did this in the past or do this in the future, it still was, or it will be, now. We humans always live in a psychological now (L. J. Hayes, 1992). If I go over there, "there" will be "here" once I get there. This is illustrated nicely in a song from the popular Swedish kids' TV series *Five Ants Are More Than Four Elephants* (something along the lines of *Sesame Street*). The lyrics go, "There is where you are not, here is here, where you are, and here is always with you." And it is in this perspective, here and now, that our experience of self, of being an "I," originates. This is because it is the perspective that always exists for us, the perspective from which everything that happens to us is experienced, and the perspective from which we learn to talk about it as "I." And because this perspective is continuous, it gives rise to our experience of continuity: the experience of always in some sense being the same person—"myself."

PERSPECTIVE TAKING AND THEORY OF MIND

The above discussion makes clear that perspective taking plays an essential role in the shaping of "self." The ability to take another's perspective is also an essential ability in its own right. In recent years it has been the focal point of extensive research based on cognitive theory, and particularly research into theory of mind (Baron-Cohen, Tager-Flusberg, & Cohen, 2000). This research aims to develop a general understanding of how humans function, how we develop the ability to attribute mental states to ourselves and to others, and how empathy arises. More specifically, this research examines the problems seen in autistic individuals and has had direct and applicable consequences, allowing development of specific strategies for helping children learn or develop perspective taking (Howlin, Baron-Cohen, & Hadwin, 1999). Based on this work, researchers have also identified what are seen as different steps in the ability to take another person's perspective. These steps start from the basic—being able to imagine what an object looks like from a position other than one's own—and go all the way to such complex behavior as being able to anticipate another's behavior from assumptions about their true versus false conceptions about events. In other words, we can act based on our assumptions about what another person believes to be true, regardless of what we ourselves believe to be true. This requires that we be able to take the perspective of the other.

Improving training methods for autistic children has occupied behavior analysts for decades. The experimental support for understanding perspective taking as relational framing opens up new possibilities in this regard. The advantage of taking on this field of research from a behavior analytic point of view is, again, that its science is built from the bottom up. The concepts that are used are precisely defined and described and have their origins in basic experimental work. On the other hand, many view it as problematic for cognitive-oriented models to define what is intended by different terms and what the basic skills are (Y. Barnes-Holmes, McHugh, & Barnes-Holmes, 2004).

In the studies based on RFT, the concepts and skills described in this book have been the starting point. Perspective taking is seen as a form of generalized operant behavior that involves having the child learn to relate stimuli from a certain perspective: I versus you, here versus there, and now versus then (Y. Barnes-Holmes, McHugh, & Barnes-Holmes, 2004). A specific protocol has been developed to train this repertoire (Y. Barnes-Holmes, Barnes-Holmes, & McHugh, 2004; McHugh, Barnes-Holmes, & Barnes-Holmes, 2009). In short, the training is designed to present the child with questions that systematically train different degrees of complexity in terms of perspective taking. Feedback is given during the training; that is, the correct answer is followed by reinforcing consequences in the dialogue.

The training starts with simple relations, such as "Which ball do I have, and which ball do you have?" and "Who is here; who is there?"

In the next step, relations are reversed: "If I were you and you were me, which ball do you have, and which ball do I have?" and "You are sitting over there in the small chair; I am sitting over here in the big chair. [This describes the actual situation.] If here was there and there was here, in which chair am I sitting, and in which chair are you sitting?"

You also train double reversed relations: "Today I am sitting here in the big chair. Yesterday I sat over there, in the small one. If here was there and there was here, and now was then and then was now, where am I sitting now? Where was I sitting then?"

As you probably noticed, answering these kinds of questions is not altogether easy, even for a verbally competent adult. That this skill corresponds to the phenomena contained in the theory of mind concept is supported by comparisons made between how this type of relational framing is manifested in groups of people of different ages, on one hand, and what is otherwise found in current research literature, on the other (McHugh et al., 2004). It has also been demonstrated that it is possible to influence this skill in preschool-aged children, through these types of training programs, and that perspective taking generalizes (Heagle & Rehfeldt, 2006). It has also been shown that autistic children's skills in perspective taking can be improved using these methods (Rehfeldt, Dillen, Ziomek, & Kowalchuck, 2007; for an up-to-date summary of this entire field, see Rehfeldt & Barnes-Holmes, 2009).

THREE ASPECTS OF OUR EXPERIENCE OF SELF

It is now time to return to what is of core interest in this book: the specific human behavior we call thinking, how it is connected to our experience of self, and how this can be understood within an RFT approach. So far I have described how RFT explains the origin of our experience of being an "I," our experience of continuity of self. You might put it this way: Not only do I see a rainbow and see my seeing a rainbow, I also see that *I* am the one seeing a rainbow; I see that I am seeing it from a specific perspective, and that a different perspective is possible. This takes place through relational framing as developed by our verbal training.

Self-as-Perspective

The perspective of continuity of self has a peculiarity. We cannot observe this perspective in itself. It can never become an object for us to observe. We can talk or write about it, just as I am doing now, and we can observe the consequences of being able to take this perspective. We can make observations from a specific perspective or locus, but we can never observe this locus or perspective as such. Of course, this is rather obvious, because from which perspective would we observe it? All we have is I-here-now. And whatever we observe, it simply cannot be this locus, as that is the vantage from which we observe it.

Consequently, our very own perspective is void of content. It is simply the point from which we observe, act, and live our lives. That is why RFT refers to this aspect of our self-experience as *self-as-perspective,* or *self-as-context.* This is the context within or perspective from which we experience what we experience. But the concept of self goes beyond this. We do not experience ourselves as being constantly void of content or as a freely floating perspective. When someone asks me about myself, I can usually describe myself in various ways, and I can also observe aspects of my experience that I would call "myself." In our attempts to analyze the phenomenological area that we describe when using the word "self," we thus need to do some more reasoning. Within RFT this is done by describing two more aspects of self: self-as-process and self-as-story. This is not to say that these three cover all of the possibilities or that other aspects of self aren't valid or useful, but for behavior analytic purposes we need these three aspects.

Self-as-Process

Self-as-process is similar to Skinner's concept of private events, which we can learn to tact. It is the ongoing, observable process of ourselves: behavior that is occurring in the moment and that makes up part of what each of us calls myself—feelings, memories, bodily sensations, and thoughts. It always exists here and now. This means that both through direct experiences and through language training, it will be connected to what the individual thinks of as "I." Note, however, that it is not constant in the way self-as-perspective is. We do not always think or remember the same thing. Our bodies do not always feel the same, and our emotional state varies.

Being firmly in touch with self-as-process is very useful. For example, what we remember and feel in many ways constitutes a summary of our history. And as described earlier, the fact that this link is useful to the people in our surroundings explains why we have learned to talk about it. How I feel right now says something about my tendency to act in different ways, and it lets people who have reason to relate to me know something about me. If I am in touch with myself as a process in the present moment, my options for acting flexibly in regard to the things that are important to me increase, and so do my options for interacting with others. If I notice that I am angry right now, I can use my awareness of this state to inform my next step. If my history has predisposed me toward aggressive behavior in a certain type of situation and I am not in touch with my own process of moving toward hitting someone, this can have troublesome consequences. In the same way, a slight affective process can be useful to me as a signal to consider pursuing an action. For example, noticing a sense of interest or curiosity in relation to some event or object can be a starting point for finding out more about it. If I don't notice this process, I run a higher risk of missing out on events that might have been valuable for me. "Knowing oneself" in the sense of self-as-process means being in touch with one's own history, including parts that were not originally verbal but now are, as a result of this very behavior.

When we notice affective states and bodily sensations in ourselves, what we are observing is to a certain extent biologically given reaction patterns (Ekman, 1992). In addition, it is partly our history of respondent learning and partly our history of relational learning. To a verbally competent person, nothing, feelings included, is simply what it is in itself. A feeling is never simply "the feeling as such"; it is also "precisely what it means to *me*." It is always a result of the individual's ability to frame relationally, and the individual's history in this respect puts its imprint on everything. Butterflies in your stomach can mean either "Oh, this is wonderful" or "This isn't going to end well." This line of discussion brings us to the border between self-as-process and the third aspect of our experience of self as described in RFT: self-as-story.

Self-as-Story, or the Conceptualized Self

From the time we take our first tottering steps as verbal beings, words will designate and refer to everything we meet, including our own behavior. We learn to talk about ourselves because this is useful to our social environment, so it has reinforced such behavior. This also means that, at an early age, we are given descriptions of ourselves that, together with our own descriptions of ourselves, develop into stories about "who I am." We learn to describe and relate this aloud, but fairly soon we also learn to do so silently. This aspect of self is termed *self-as-story* or *conceptualized self.*

Words and thoughts rapidly invade our experience for many reasons, among them that verbal behavior can occur simultaneously with more or less any other behavior (Parrott, 1984). In general, it is difficult to carry out several nonverbal actions simultaneously with the same object, but it is easy to think of something while at the same time carrying out other actions with the same thing. It is easier to think of a chair while painting it compared to standing on it to reach the ceiling while simultaneously painting the chair. This is so obvious that we normally do not think about it. But the very fact that verbal behavior usually does not have a direct effect on our physical environment makes it extremely common for verbal behavior to proceed simultaneously with almost anything we do. Part of this verbal behavior describes and relates our own actions.

A story or conceptualization of ourselves is of course very useful. It contributes to continuity and supplies the individual with a type of summary of "who I am." This sort of summary is essential in our interactions with others. Because it can be presented, it can, to a certain degree, stand in for direct experience of an individual by his social community. It can summarize his history, what he thinks is important, what he can be expected to do, and so on. For this individual's social community, this story becomes a shortcut to knowledge about him. This hearkens back to the above discussion of the function of learning to talk about ourselves as processes. Self-as-story is, however, much more verbally elaborate than self-as-process, while also possibly omitting a lot of what is going on with the individual in the moment. A typically formalized way of using self-as-story is the type of summary we present in a job interview or when we introduce ourselves in a new social environment.

Think of how much of our interaction with children involves teaching them to create this story. We ask questions and make statements like these: "What did you do then?" "What did you think about that?" "Are you a boy or a girl?" "You look so much like your father." "You are so cute." These questions and statements, along with the dialogues that follow, help children form a conceptualized self. An important part of this story is that it is coherent and a connected whole. This becomes evident when the individual does things that don't seem to resonate with the already existing story of "me" in a certain context. For example, if someone whose self-as-story includes being kind and forthcoming loses his temper in a certain situation and rejects someone's request, this can lead to the person telling himself and others things like "I wasn't myself." Something similar occurs if someone says something about us that doesn't correspond to our own story. We look for explanations and ways of formulating the story to make it coherent. We defend our story about ourselves, and if we do change it, we do so in a way that makes it stay logically consistent.

This brings us back to what I wrote earlier about coherence as a generalized reinforcer. The conceptualized self, which has its origins in our interactions with our social environment, soon becomes important to us. And before long, parts of these stories will be known only by the individual, since we learn to tell them silently and thus, for different reasons, can withhold parts of them from our environment. There are aspects of self-as-story that we don't relate in job interviews. We probably tell a different version of our story to our friends and closest family members, but this information is also selective. Some aspects of the story we tell about ourselves remain known only by the individual. In many ways, our stories about ourselves have the same function for ourselves as they do for our environment. They turn into summaries that, among other things, function as scripts for action (Tomkins, 1987): "I am a person who does this or that." This leads us closer to the topic of the next chapter: rule-governed behavior, and self-rules in particular.

Self-as-story entails several obvious risks, just like all summaries do. First of all, this self-conceptualization is, by necessity, an extreme simplification. This is its very point and at the same time a limitation. In light of everything that has happened in my life, my conceptualized self is just one of many possible stories. Parts of the story have been created as a result of my own tacting of self-as-process: what I have noticed about my own behavior, feelings, memories, bodily sensations, and thoughts. At the same time, it has come about as a result of my interaction with a social environment that, through its tacting of my behavior, has also created parts of the story. We can assume that at least some parts of this tacting have been what Skinner referred to as impure, or distorted (Skinner, 1957); in other words, parts of my story of myself have been influenced by things other than my behavior. In more everyday language, versions of my story of me have been created by others, and many times they have had purposes other than simply describing me when telling the story. A leading psychoanalyst has expressed this very tellingly: You could be "knowing what you are not supposed to know and feeling what you are not supposed to feel" (Bowlby, 1988, p. 99). This ties in with another important quality in our self-conceptualization. A core part of the story consists of comparative, evaluative framing. Self-as-story always contains what is good and what is bad with me specifically. And here lurks the deepest fear for all of us: that we are fundamentally bad, aversive, and not the way we "should" be.

It is easy to see that when these different limitations in self-as-story are combined with our efforts to keep that story whole and coherent, the story can become a pitfall, rather than an asset. This is something I'll return to in later discussions on clinical issues, in both chapter 7 and part 3 of the book.

THE RELATIONSHIP BETWEEN THE THREE ASPECTS OF SELF

The preceding survey of three aspects of self reveals that the distinctions between them are not always clear. However, sharply delineated distinctions are not necessary. This is not an attempt to describe the three parts of self as if they actually existed. This is just a way of speaking about our experience and relating it to the scientific data that are summarized in RFT. The main point is that our capacity for arbitrarily applicable relational responding is what gives us the experience of these different aspects of self, and that this is significant in the clinical application of this knowledge.

That said, there are a few things that may be worth mentioning about the relationship between these three aspects of self. First a word on an important distinction between self-as-perspective and the two other aspects. Self-as-perspective is the position here and now, from which everything is observed, and is not itself open to direct observation, whereas both self-as-process and self-as-story are aspects we can observe directly. This creates a potential experiential distance between self-as-perspective, on one hand, and self-as-process and self-as-story, on the other. Yet this experience of distance is often missing. A person can easily coordinate self-as-process and self-as-story with I-here-now; that is, act as if these aspects define him. But we have the possibility of experiencing both self-as-process and self-as-story as being there and then, as when the individual discriminates something he feels or thinks as what typically turns up for him, for example. This also means it is possible to take a different approach toward a restricting story about oneself. A direct experience of the experiential distance between I-here-now and I-there-then can lead the individual to the position of "That is a story about me that life has given me, and—being able to observe it—I am more than my story about myself." This, of course, brings out a new story, but because the new story will appear as a result of new experiences, it could be more useful and less restricting. I will discuss this at greater length in part 3 of the book.

As implied above, self-as-process and self-as-story are similar in the sense that both constitute "content of self," which can be observed there-then. Among the regards in which they differ is that self-as-process is connected to more direct, nonderived stimulus functions, while self-as-story is more dominated by derived stimulus functions. To use an alternative expression, self-as-story is more cognitive. Another way of describing the difference is that self-as-process focuses more on the present, while self-as-story is more abstract and static and embraces the individual's overall history to a higher degree. Another way of describing these two aspects is as self-as-process and self-as-product, respectively.

SUMMARY

This chapter described how basic relational framing provides the building blocks for more complex relational behaviors such as relating networks of relations in analogy and metaphor. It also described how perspective taking forms the foundation of our experience of self, which manifests in several ways. All of this is central to human behavior in general, and thereby also key to what we often refer to as "psychopathology." Next we'll turn to what Skinner referred to as rule-governed behavior. This is the area in which verbal behavior has its most far-reaching consequences for the conditions under which we humans live. It is also in connection with this ability that the side effects of language become most evident. We will get to those side effects in chapter 7. But, first, let's take a closer look at rule-governed behavior in itself.

CHAPTER 6

Relational Framing and Rule-Governed Behavior

The term "rule-governed behavior" was first used by Skinner in a chapter on problem solving (1966). As mentioned in chapter 2, since then behavior analysts have struggled to describe this type of complex human behavior. My intention in this chapter is to show how rule-governed behavior can be analyzed through RFT and how this opens up prospects for a better understanding of this phenomenon. I will then describe different types of rule-governed behavior and discuss how they develop and how they are important to human behavior.

The concept of rule-governed behavior is based on functional analysis. A certain behavior is understood and influenced by analyzing contextual factors: what precedes (antecedent) and what follows upon (consequence) that behavior. In an attempt to describe rule-governed behavior based on this paradigm, Skinner discussed how certain antecedents function as rules or instructions. They specify behavior and consequence. You might say they function as if they anticipate what has not yet taken place. Another important aspect is that they prescribe behavior. A child is told, "Put your jacket on and you'll stay warm," and based on this statement she puts her jacket on. Someone says, "If you want to convince her of your love, you have to spend more time with her," and the person who hears this makes some changes in his time priorities. We give these kinds of instructions to each other and ourselves almost continuously.

We could summarize the problem with traditional functional analysis of this behavior, described in chapter 2, in this way: How can antecedents acquire stimulus functions from something that, seemingly, is experienced as existing in the future, or from other events that the individual has not been in contact with? And how can an antecedent influence behavior that occurs much later—maybe years later—although it did not have this effect on behavior when the antecedent was present? Here's an example: Someone tells a traveler, "When you visit Stockholm, you should go to the Vasa Museum." When the traveler finally visits Stockholm many years later, what makes her choose to go to this museum based on what was said so long ago? If we place the reasons for this in an internal world of mental representations, we might feel that the problem is solved. But, as outlined earlier, within behavior analysis this answer has never been good enough.

We can answer these questions through relational frame theory, and we can do so while staying true to the basic assumptions of behavior analysis. Early in life, humans learn a generalized operant: arbitrarily applicable relational responding. This responding is governed by contextual cues that specify the relation so that the relational response can be brought to bear on any stimuli, regardless of their formal, physical properties. This relating will, in turn, govern which stimulus functions are cued in a given moment. After someone tells me that many people in the neighborhood are suffering from an upset stomach because of eating poorly prepared chicken, eating the chicken on my plate will be in a relation of coordination to something aversive: becoming sick. In this situation, the chicken acquires stimulus functions for me that it did not have before I heard this series of sounds ("...many people in the neighborhood..."). This response does not depend on my having previously become sick after eating chicken. I may never have experienced stomach illness at all.

Note that this does not occur because of some hidden process inside the person. Of course there are things occurring inside the person, just as in any other behavior. However, what I have described here is a behavior performed by the person as a whole. Humans relate things in this way, in what can perhaps be most accurately described as a social game. This is the key to understanding rule-governed behavior.

RELATIONAL FRAMING AND THE TRANSFORMATION OF STIMULUS FUNCTIONS OF ANTECEDENTS

Relational framing alters the traditional ABC sequence because it influences how the components of the sequence acquire their functions. In a traditional sequence, the antecedents, for example, obtain their stimulus functions by direct contingencies in the earlier history of the organism or according to physical properties, as in generalization. If a dog is passing a certain gate and another dog attacks it, it is likely that this gate will acquire new stimulus functions for the dog that was attacked. From that point on, the gate may have the stimulus function of signaling danger, due to respondent learning. Other gates could take on similar functions through generalization, provided that they are similar enough. Similar learning occurs in humans, but we also have another possible path to learning, through arbitrarily applicable relations between stimuli. A gate can acquire new stimulus functions for me without my having any direct experience of this gate or anything that is similar to the gate in any way. All that is required is for another person to utter a series of sounds that are arbitrarily agreed upon: "Don't go in there, or you might be attacked by a fierce dog." A sign with a series of black characters on a white background spelling out "Beware of dog" can have the same function for my behavior, even if I've never encountered a sign like that before—by this gate or anywhere else. That both the sounds and the visual stimuli are arbitrary is easily illustrated by the fact that they probably wouldn't have the same functions for someone who is only familiar with the Japanese language, for example.

For an antecedent to function as a rule in this way, a certain skill is necessary in the listener: being able to relate stimuli in coordination, so that the different parts of the rule—the sounds and the words—stand for something. In this case, the words "fierce dog" are put in a relation of coordination with an actual fierce dog. If the rule is to be meaningful and understandable, it is also necessary for the listener to relate stimuli temporally and causally, to establish the relation between the behavior and its consequences, which are either described or implicit in the rule. In this example, the listener needs to be able to put the behavior of "going in there" in a temporal and causal relation with the consequence "might be attacked."

Note that rules or parts of rules can be implicit (D. Barnes-Holmes et al., 2006). In contrast to the way we traditionally conceive of rules, not everything that alters stimulus functions must be explicitly expressed in the rule. Here is a classical verbal antecedent functioning as a rule: "Be careful and you'll make it!" The rule specifies behavior (be careful) and consequence (you'll make it). This could be followed by the listener being extremely careful about revealing personal information in a particular situation, for example. But simply uttering "Be careful!" can be followed by the same rule-governed behavior, even though this statement doesn't seem to specify any consequences. The same rule-governed cautious behavior could also follow upon simply seeing another person act in a certain way, without anything being uttered. And the person in question can follow this rule as though it is generally valid, that is, valid essentially all the time. This is an example of how rule-governed behavior can follow implicit rules, which, according to RFT, is easily explained: The rule that is followed is not necessarily the rule that was stated. The rule that is followed is the rule that was contacted. And which rule is contacted is determined not only by what was said, but also by the listener's learning history, both direct and derived. This learning history gives the circumstances that are present a specific function and thereby influences the individual's actions. My history could be such that merely being in the presence of other human beings puts me in contact with "Be careful and you'll make it!"

So the question of how a rule can specify behavior and consequences that are not current and that the person has not earlier experienced (Schlinger, 1990) is answered by invoking arbitrarily applicable relational responding (D. Barnes-Holmes, Hayes, Dymond, & O'Hora, 2001; O'Hora, Barnes-Holmes, Roche, & Smeets, 2004). A rule puts the listener in contact with a relational network[1] that transforms the functions of the stimuli that are related to the network. At her actual visit to Stockholm many years after the rule about going to the museum was stated, the current circumstances have certain stimulus functions for the traveler—functions they wouldn't have had if the rule had not been uttered and the traveler had not related and did not now relate the present Stockholm to a certain museum. Going to the Vasa Museum while in Stockholm can, of course, be governed by completely different factors. But if, in this case, it is a result of the rule that was stated many years ago, then this happens because present conditions have acquired their stimulus functions through the social game we learned to take part in when we learned to relate events arbitrarily.

RULES CAN BE UNDERSTOOD WITHOUT BEING FOLLOWED

It is worth noting that even if a rule is heard and understood, it is not necessarily followed by rule-governed behavior. It can be understood and followed,

1 The reader is once again reminded that the use of the term "relational network" does not imply that any such objects exist. To talk about relational networks is to say that humans act in a particular way: relating events in a potentially complex way. This way of relating affects the stimulus functions of such events.

or it can be understood without being followed. Understanding is simply the ability to relate stimuli in the way discussed above. Several factors may influence whether a rule is followed or not.

For example, the specific behavioral repertoire needed to follow the rule may be missing. You may completely understand the rule "Hit the dead center of the bull's-eye and you'll win the shooting competition," but if you cannot handle the gun you are to use, you will not follow this rule.

Another example of a factor that influences whether rule-governed behavior occurs is the credibility of the person uttering the rule, as seen from the listener's perspective. Most readers of this book probably wouldn't follow a rule on how to carry out psychological treatment if it was uttered by an actor referring to the signs of the zodiac. However, if the same rule was delivered by a prominent psychotherapy researcher and she was referring to new scientific findings, it is considerably more likely that the rule would result in rule-governed behavior. The speaker's credibility can be based on the listener's actual experiences of following rules uttered by this speaker, or on derived stimulus functions. An example of the former would be following a piece of advice from a lifelong partner or close friend who has earlier given counsel that was helpful. An example of the latter is the way in which we normally follow rules given by various experts or, for that matter, when we do not follow this type of rule because we follow the rule "So-called experts are usually wrong."

In certain situations, we may also find that contingencies of reinforcement for rule-governed behavior are missing. Here's an example: A child may follow her parents' rules but not her sister's, since all of the reinforcement she's received so far for following rules has been connected with her parents, not her sister. The same effect could also be the result of what was described above as lack of credibility. The only way to determine which is the case is by analyzing the specific situation.

Yet another reason why a rule may not be followed even though it is understood is if the rule is incoherent or contradictory in relation to the listener's learning history. This is what, in everyday language, we might describe as obviously not correct. Here's an example: "Stay as sedentary as possible, smoke at least twenty cigarettes a day, and regularly use large amounts of alcohol, and you will increase your chances of a long life and good health." Very few of us, if any, would act on a rule like this.

DIFFERENT TYPES OF RULE-GOVERNED BEHAVIOR

There are two basic forms of rule-governed behavior, distinguished by the type of history of reinforcing contingencies associated with them.[2] These are called pliance and tracking. A third form of rule-governed behavior, called augmenting, works in combination with either of the other two by influencing the degree to which the consequences specified in the rule function as reinforcing or punishing.

Pliance

Pliance is "rule-governed behavior under the control of a history of socially mediated reinforcement for coordination between behavior and the antecedent verbal stimuli (i.e., the relational network or rule), in which that reinforcement is itself delivered based on a frame of coordination between the rule and behavior" (S. C. Hayes, Barnes-Holmes, & Roche, 2001, p. 108). It is similar to what we mean by "doing as you are told," as said in everyday language, because it implies having previously encountered reinforcing contingencies that promoted doing precisely this. The behavior of following the rule, in and of itself, is what matters in pliance, since the consequences are controlled by the rule giver and are dependent on following the rule. Typical pliance is when one person yields to what someone else says in order to obtain that person's approval, provided that this is done based on the consequences that are specified in the rule. If I am stopped by the police and asked to show my driver's license, doing so is probably an example of pliance. The rule that precedes pliance is called a *ply*. When the rule is stated, this behavior by the speaker is an example of the type of verbal operant that Skinner called a mand (described in chapter 2).

The governing consequences are, of course, only apparently contacted through the rule—this being an antecedent. The subsequent behavior has not yet encountered its actual consequences. In rule-governed behavior the consequences specified in the rule need not have been contacted by the listener at a previous point, distinguishing this behavior from that governed by direct

---

2 This functional distinction was described briefly by Skinner (1966), though he did not use the more elaborate terminology used here. And as mentioned earlier, he did not give a detailed analysis of any learning history that makes this kind of behavior possible.

contingencies. However, the listener has previously encountered the direct consequences of following rules as such. A verbally competent person can follow a rule that specifies a consequence like "or I'll shoot you" without ever having encountered such a consequence before. What is necessary is a repertoire of arbitrarily applicable relational responding, combined with having previously encountered consequences of following rules. By definition, this applies to all forms of rule-governed behavior.

How do we learn pliance? Imagine a small child doing something that is controlled by immediate consequences, like touching something interesting, say her mother's new laptop. Her mother may want this behavior to stop. This can, of course, be achieved by manipulating direct consequences, such as by removing the laptop or taking the child to a different room. When the child, through language training, learns relational framing, sounds may start to function as verbal stimuli. These stimuli, by their presence, alter the function of the present circumstances by relating these circumstances arbitrarily. A child with a particular history of relational framing will not touch the laptop when she hears rules such as these: "Don't touch Mommy's laptop, or I'll make you leave the room" or "If you don't touch Mommy's laptop, I'll give you a nice surprise later." In this context, the laptop has been transformed from something to touch into something related to the consequence "leaving the room" or "a nice surprise later." Rule-governed behavior is reinforced by similar rules being stated on a large number of occasions and being followed by socially mediated consequences specified in the rule.

It is easy to see the enormous advantages that arise with this type of influence on human behavior. One advantage is that new consequences can be added by the social environment. Another is that remote consequences can be contacted, which may override more immediate consequences since both behavior and consequences are specified in rules. And all of this can occur without the individual contacting the consequences directly. A child can be made to refrain from playing with interesting objects by being put in verbal contact with aversive consequences of that type of playing, appetitive consequences of refraining, or both. New members of the human herd continuously learn to be influenced by remote consequences that are specified in rules issued by the rest of the herd. Of course, this requires the appropriate relational framing of coordination, comparison, causality, conditionality, and so on. This is where the secret of our ability to respond to delayed contingencies and override immediate consequences lies. Pliance is the first type of rule-governed behavior that we learn. Then, based on this skill, we develop tracking.

Tracking

Tracking is "rule-governed behavior under the control of a history of coordination between the rule and the way the environment is arranged independently of the delivery of the rule" (S. C. Hayes, Barnes-Holmes, & Roche, 2001, p. 109). A typical instance of tracking would be the behavior of someone driving in a certain direction after hearing, "Go straight ahead for about half a mile, then make a right turn when you see a gas station, and two hundred yards ahead is the sports field." This example is valid provided that the person drives as directed under the influence of the apparent correspondence between the rule and the factual location of the sports field—apparent when hearing the rule uttered, that is. After all, in this case it is the rule that governs the behavior, not the actual location of the sports field. If the rule is followed by the listener, it functions as a *track*. As a verbal operant of the speaker, it is an example of a tact (see chapter 2).

Tracking is taught by the social community after a certain degree of pliance is in place. Let's go back to the example of the child and the laptop. When a young member of the human herd, through pliance, can override immediate consequences (like the rewarding effect of touching her mother's laptop), she will go on to contact other available consequences. These consequences are not necessarily socially mediated; they are a result of how the environment is arranged, and she would not have contacted them if immediate consequences had still been dominating her behavior. In the learning situation, this may occur because the social community arranges for the child to contact the consequences this way, or simply because everything is constantly changing. If the child were to stay close to the laptop without touching it in an instance of pliance, then at a minimum she will contact "the laptop when it is not touched." Let's say that the laptop was just about to display a sequence of interesting pictures. If she indeed does not touch the laptop, she will encounter these pictures as a consequence of not touching. These consequences may now in turn be specified by the social community, provided that the relevant training of relational framing has occurred. Rules can now be formulated that seemingly put the child in contact with these consequences, and her behavior can thereby be influenced via these very rules. This means the young member of the herd goes from being able to act on rules that specify consequences placed there by the social community to being able to act on rules that put her in apparent or indirect contact with all kinds of events.

Let me give you a perhaps more likely example of how the social community arranges for this learning to take place. When a child has finished playing, her father might say, "Look how dirty your hands are. Let's go and wash them to make them clean again." Let's assume that the child comes along due to pliance. She follows the rule based on a history of reinforcement that can be described as "You're supposed to do what Daddy tells you." When the child's hands are washed, her father might remark on how clean her hands are without adding any social consequences as a result of the girl doing as she was told. At this point, the father can help his daughter discriminate the changes that take place and provide relevant relational training, like framing events in terms of coordinate, temporal, and causal relations. He could say, "Look, your hands are really dirty. Rinse them with water, and look—what happened?" He could ask the child what she did, what happened then, and why it happened. (For a more thorough account of this kind of training, see Luciano, Valdivia-Salas, et al., 2009.) The social community provides the child with many samples of rules that specify behavior and actual consequences that are reinforcing or punishing in themselves, independent of the socially mediated consequences that are dependent on rule following as such. This will gradually make it more likely that the child tracks further rules. Initially this will at least apply to interactions with people who are important to the child. This is the starting point of what I described above as the speaker's credibility.

Augmenting

Augmenting is "rule-governed behavior due to relational networks that alter the degree to which events function as consequences" (S. C. Hayes, Barnes-Holmes, & Roche, 2001, p. 109). Augmenting occurs in conjunction with either pliance or tracking.

How augmenting occurs is explained by RFT as follows: A relational network is related to a consequence and thereby alters the strength or function of this consequence. Let me return to the girl who has learned to wash her hands. "Use the green soap, Maria, and your hands will be clean." This rule can be followed by the girl washing her hands with the soap as an instance of pliance. In this case the behavior occurs because, for Maria, the rule implies consequences of following rules as such. It could also be that Maria has learned tracking, and that she uses the soap based on the specification that her fingers will be clean. But if the rule she follows is "If you use the soap, you're a smart girl," this might be an example of augmenting. This is the case if being "a smart girl" has a reinforcing function in itself. Augmentals are rules that are not restricted to specifying a consequence that has not been contacted but will be (like the location of the sports field or clean hands); they also allow us to contact consequences that are abstract and can exert influence over behavior without ever being contacted directly. For example, a person can act on rules that specify consequences after death, which, by definition, no living person has contacted. Likewise, we can act based on consequences that are too abstract to be available for direct contact, such as "an equitable international economic order."

Note that this type of rule-governed behavior is connected to either pliance or tracking. Augmenting can be described as a separate unit, but the way in which it exercises its function is by influencing tracking and pliance through altering the reinforcing or punishing qualities of the specified consequences.

Two types of augmenting are described in the literature: formative augmenting and motivative augmenting. *Formative augmenting* is behavior due to a rule that establishes a given consequence as reinforcing or punishing. A formative augmental, then, gives reinforcing or punishing qualities to some outcome that previously did not have these qualities by relating it to an already established reinforcer. A formative augmental creates a motivator, so to speak. Let's say someone sees a worn *Donald Duck* magazine in a flea market but doesn't have any special interest in old comic books. Then someone else says, "Hey, that's the very first issue of *Donald Duck* magazine. It's a rarity." For the listener, this statement can function as a formative augmental that makes it more likely that she'll buy the magazine. "Very first issue" and "a rarity" are already verbally established reinforcers. By being related to them, the worn magazine becomes a reinforcer as well. A rule (in this case a track) like "Buy this and you will be the owner of a rarity" might affect this person's behavior. Another example occurs when a man is introduced to someone who doesn't strike him as particularly interesting—that is, not until someone tells him that the man is the brother of a woman who does interest him. The statement may then function as a formative augmental for his further actions in relation to this person. His contact with this person has acquired new worth, and with it a higher probability that his actions will be influenced by a track like "Stay close to this guy, and you will be close to Barbara's brother," as the formative augmental suddenly puts him in contact with a previously established reinforcer.[3] Speaking to this unknown person has become reinforcing through a formative augmental.

3 The observant reader probably notices the similarity between the process described and generalization. Note, however, that generalization requires that there exist a formal similarity between stimuli, or that the stimulus that acquires its function by generalization has been contingent with a primary reinforcer. This is not the case here; the stimulus functions are altered through arbitrarily applicable relational responding.

Motivative augmenting is behavior due to a rule that alters the probability that an already established reinforcer (or punisher) will function as reinforcing (or punishing) in a certain situation. A motivative augmental highlights a motivator, so to speak. Let's say that someone is already interested in old comic books. He is a collector. One day he is skimming through his daily paper and sees an ad for his local antiquarian bookshop: Copeland's Antiquarian Bookshop—Books and Comic Books. He already knows about the bookshop. In fact, he goes there regularly. He does not need the ad to become aware that they sell comic books. But the moment he reads the ad, the shop seems more important to him; the ad puts him in emotional and sensory contact with an already established reinforcer, bringing it to the fore in the present moment. This increases the probability that a rule like "Take a stroll over there today and see if you find something interesting" will become governing. If he acts on this because he saw the ad, it would be an example of motivative augmenting. Another example is a situation where a father is loaded with work one day and that evening there is a parent-teacher meeting at the school his eight-year-old son attends. When speaking to a coworker, he says that he actually wants to go, but he doesn't have time; there's too much to do at work. His colleague replies, "Well, you usually say it's important to you to be a dad who's there for your kids." If this makes him go to the meeting after all, based on the rule "Go to be there for your son," despite the fact that he had been prepared to give it up, then his coworker's reminder functioned as a motivative augmental.

Both types of augmenting influence what we commonly call motivation: how important things seem to us. A formative augmental establishes something new as reinforcing, and a motivative augmental temporarily increases the reinforcing value of something that is already reinforcing. In everyday language, you would say that a formative augmental makes something new important, and a motivative augmental makes something that is already important even more important in the moment. This process can function in the opposite direction, as well. Something that functions as a reinforcer can lose its strength or entirely cease to be reinforcing through augmenting. If an intrigued comic book collector has just found a rarity and he then hears someone say, "But it's damaged; it's missing the most important page," then the likelihood that he'll buy it probably decreases. This is provided that the person who said this was credible, in accordance with the discussion earlier in this chapter. An analogous example would be if a fellow was on his way to see some people, one of them being Barbara, the woman he's interested in. Then someone says, "Barbara is bringing her new boyfriend, Steven." If this makes him not go, the get-together has lost some of its apparent reinforcing function through augmenting. (Again, the point of using the word "apparent" here is that the actual get-together has not taken place. Its prospective reinforcing function is yet unknown. Here I am describing antecedents that function as rules.)

This dynamic can be likened to the phenomena referred to as establishing or motivational operations within behavior analysis (Michael, 1993). These are antecedents that influence the degree to which a given consequence functions as reinforcing or punishing (see chapter 1). This corresponds to our everyday way of speaking about it too. The comic book collector in the example above could very well describe his interest in comic books in terms of a hunger or a thirst. For example, he could say something like "My appetite was whetted to check whether they had something new in the shop" as a way of explaining how the ad resulted in a visit to the bookshop. The relational networks or rules I have described using the term "augmental" thus function as verbal establishing operations (S. C. Hayes, Zettle, & Rosenfarb, 1989).

The fact that augmenting is connected to what we commonly call motivation makes it clear that this form of rule following is central to human behavior as a whole. What motivates us in life is crucial to how we live and behave. Augmentals are the relational networks that put us into contact with the things we value in life, the things that are truly important to us. In the example above, going to the parent-teacher meeting was motivated based on "being a dad who is there for his kids." We can assume that, for this father, this functions as a motivative augmental for a number of different decisions and actions. Many everyday events or processes, in themselves neutral—or even boring or painful—obtain their reinforcing functions based on verbal establishing operations. Sitting in front of the computer to do my writing when the sun is shining outside and it is one of our first warm summer days in over a month—the way I am doing as I write this—is not very rewarding in itself. I am doing it based on augmentals—based on the bigger picture of the purpose of my actions. Whether explicitly or tacitly, our lives are based on certain assumptions about what we want our lives to stand for or be all about. This is what we commonly refer to as values. These are verbally constructed consequences that are globally desired by the individual and function to help us determine overall directions in life. They are created through relational framing and, as a result, can be present and influence our actions in a large number of situations. One area where augmenting is at the very core of behavior is in what we commonly call moral or ethical behavior (S. C. Hayes, Gifford, & Hayes, 1998).

The fact that this type of rule following is of such significance in our lives also means that it is highly relevant to clinical problems. This is hardly news. Values have long been the focal point of different psychologies and philosophies (Leigland, 2005; Dahl, Plumb, Stewart, & Lundgren, 2009). Hopefully, an RFT-based scientific analysis of this area will further contribute to the development of helpful clinical interventions. I will return to this in part 3 of the book.

Rule Following Is Defined Functionally

Let me finish this description of the different types of rule-governed behavior by emphasizing, once again, that these are functional units. The type of reinforcement contingency that governs the relevant behavior determines which kind of rule-governed behavior is at hand in a given moment. The formal topography of the rule is not decisive. Let's return to the man who collected comic books. Let's say that it was not he himself, but his wife, who caught sight of the ad. She tells him, "If you go down to the antiquarian bookshop today, now that you have a day off, you might find some of the magazines you want." This can function as a rule that influences his actions in the exact same way as if he had seen the ad himself. In that case, her words function as a track combined with an augmental. But her words can function as something else, even if the rule is stated in exactly the same way. The husband might go down to the bookshop because he wants to please his wife, as he thinks that this is what she wants him to do. If this is the situation—that he wants her approval—and he goes to the bookshop because of earlier consequences of following rules as such, then what his wife said functions as a ply and his behavior is pliance. The function of the statement for his wife is irrelevant to what type of rule following his actions constitute.[4] What determines the type of rule-governed behavior he engages in is the rule he is in contact with and that he acts on, nothing else.

Pliance can, of course, be performed simply based on the individual's experience of following rules—on the fact that doing so has previously been reinforcing. This is the case early in our learning history, as described above. But pliance is also affected by augmenting. If the comic book collector acts in order to obtain his wife's approval, he might do so based on verbally constructed consequences beyond being approved of in the moment. He may do it based on his assumptions about how a husband is supposed to behave, or

---

4 Formally her statement is a tact; that is, it is governed by what precedes her statement—the ad. It could, however, have been a false tact and actually constitute a mand if she asks her husband to leave for a while and her statement is governed by earlier consequences of asking him to do so. In that case, it's just that the topography she's using makes it look like a tact. As a reminder, the speaker's verbal behavior is also functionally defined, as described in chapter 2.

possibly based on even more encompassing values concerning what is important in interactions with other people. This provides another example of how defining augmenting is for human behavior in general.

Let's take another look at something I wrote earlier: that rules can be implicit. Let's assume that the man in this example went to visit Copeland's Antiquarian Bookshop. He may have done this due to pliance: He acted based on the rule he contacted because of what his wife said, a rule that could be described as "If I leave the house for a while, she will be pleased." In this case, he did what he did in order to please his wife.

Yet this is not what his wife said, so how can we identify this rule? If we were to ask the husband, he might tell us that this was his train of thought as he chose to go out for a while. But that need not be the case. He might not have had the experience of thinking anything in particular. Still, to him, the rule or instruction might have been implied in what his wife said, so his behavior was rule-governed; he acted on the rule he contacted.[5] But in that case, where was the rule? At this point, we are approaching the phenomenon that in psychodynamic theory is explained using the concept of the subconscious. Within cognitive theory the reference is usually to another postulated inner phenomenon: schemas (Beck, 1964; Young, 1990). If we describe this type of phenomenon as a following of implicit rules, we may ask ourselves, where is the rule itself? However, from a functional contextual perspective, the rule need not exist as a concrete phenomenon. It may, if it is stated or thought by someone. If it is not stated and is not in someone's thoughts, but simply implicit, you might say it exists in the interaction of the moment, in the interaction of context and response (L. J. Hayes, 1992). What distinguishes the behavior as verbal rather than simply governed by direct contingencies is the way in which this interaction takes place. If relational framing is involved, then as defined by RFT, the behavior is verbal. If the rule is subsequently formulated, its content is just a verbal abstraction of the behavior. This, then, is yet another example of relational framing.

Experimental support for the phenomenon of rule-governed behavior as understood from the perspective of RFT is not as elaborate as support for the basic phenomenon of arbitrarily applicable relational responding. However, several relevant studies are available from recent years (O'Hora, Barnes-

5 It is also possible that his behavior was governed by direct contingencies. Again, we are up against the difficulties with everyday examples. We would have to know the individual's learning history to be certain to what extent a behavior is governed by direct contingencies versus rules.

Holmes, & Roche, 2001; Whelan & Barnes-Holmes, 2004b; Whelan, Barnes-Holmes, & Dymond, 2006; Ju & Hayes, 2008).

SELF-RULES

Self-rules are rules directed toward oneself that influence one's own actions. Taking into account the analysis thus far in this book, self-rules therefore require both a certain level of competence in rule following and an experience of self along the lines of the aspects of self discussed in chapter 5. A core characteristic of the behavior of following self-rules is the same as for rules given by others: A certain behavior follows, controlled by the apparent consequences specified by the rule, rather than by direct contingencies alone. As I've emphasized, this is the very foundation of the human capability that might be summed up as delayed responding. Self-rules can be relatively simple, like "If I hurry, I'll be able to catch the bus," or more complex, like "If I can just get rid of my anxiety, I'll be able to do what I want in life."

The ability to lay down rules for oneself is consistent with what I've outlined about a growing experience of self and the ability to follow rules given by others, and in principle there is no need to add anything to this. On the contrary, given these abilities you might say that developing self-rules is inevitable (Luciano, Valdivia-Salas, et al., 2009). We could describe the sequence in the following way.

A child learns, through direct contingencies of reinforcement, to tact her own behavior, including private events like thoughts and feelings, as described in chapter 2.

Relational framing increases the complexity of this behavior, since different phenomena can be related arbitrarily in line with the training the child receives from the social community.

Parts of this training help the child develop increasingly complex tacting of "me" and, in tandem, gradually acquire the three aspects of self described as self-as-perspective, self-as-process, and self-as-story.

The child can now observe herself as an object of other people's actions and her own actions.

Alongside this, the child learns rule-governed behavior. The process starts with rules provided by others: "Mary, eat your food." Early on, this utterance will probably be echoed by the child. After the first basic ability of relational framing (coordination) is in place, the echoic behavior of the child, "Mary eat food," can be transformed to "I eat food." A next step is a rule such as "If you [Mary/I] eat the food, we can watch TV afterward." Pliance is followed by tracking and augmenting. The different words that are a part of the training the child receives in relating stimuli arbitrarily will gradually become a part of various relational frames, and so the child's behavioral flexibility increases. Whatever you can say aloud, you can also learn to say silently. So self-rules evolve alongside both the ability to engage in rule-governed behavior in general and successively more complex experiences of different aspects of "me."

Imagine a preschool-age boy who has just been told that the person he has always called Mother is not his real mother, but that she is his younger siblings' real mother. If "mother" is in a relation of coordination to experiences like security, joy, and a number of other things that are important to the boy, he might suddenly—simply based on the negation "is not"—derive thoughts about not having these things. He could derive thoughts about his mother leaving him, provided that, for the boy, "mother" is in a relation of coordination to the experience of the mother being there for him, and under the condition that he has the skill of framing temporally. He may also derive thoughts about his younger siblings and his relationship to them—maybe thoughts about being different from them. This ability to relate events, together with a number of things that are actually going on in the situation, can become a part of the story about who he is. It is easy to see some self-rules that may result. Since what he has been told has put him in apparent contact with a number of events that are painful to him, it is possible that he will want to avoid experiencing this again. This can lead to a self-rule like "Don't talk about this," since talking about it will necessarily put him back in contact with this pain. Note that I am using the word "apparent" again. It is clear that the boy can be experiencing considerable suffering. And yet he has not actually encountered any of the possibilities he is deriving or that scare him. All he has encountered is a series of his own responses, what we call "thinking." This may seem obvious to us, since we are all in the same social game, but it is actually a remarkable thing.

This discussion hints at some of the consequences of the capability for arbitrarily applicable relational responding and rule-governed behavior that are essential to clinical problems and clinical work. (I will return to this in part 3 of the book.) Nonetheless, we should not let this overshadow the fact that the ability to follow self-rules first and foremost increases our behavioral flexibility. We can tell ourselves to keep studying, even when it is taxing and anything but rewarding, in order to pass our exams and be able to work in the field of our choice. We can talk to ourselves about things we have never experienced and direct our actions in a way that increases the likelihood of actually achieving something previously unexperienced. We can hold on to our ideals and orient our actions and lives toward accomplishing long-term goals that serve both ourselves and others.

RULE-GOVERNED BEHAVIOR AND PROBLEM SOLVING

As mentioned at the beginning of this chapter, Skinner first used the term "rule-governed behavior" in a chapter on problem solving (Skinner, 1966). He had written on the topic of problem solving earlier, as well: "We are concerned here with the process of 'finding the solution.' Problem solving may be defined as any behavior which, through the manipulation of variables, makes the appearance of a solution more probable" (Skinner, 1953, p. 247).

According to this definition, not all problem solving is verbal. If I try to unlock a door and the key does not quite fit, I may try to adjust it in different ways and thereby manage to open the door. My behavior alters variables, in this case the key's relation to different parts of the lock, and this is possible without rule-governed behavior. But rule following may be involved—if I am telling myself what to do, if I think back to what I might have done earlier in order to solve this problem, and so on. The latter is the type of problem solving of interest in this book. The exact definition is not important in this context, as RFT does not deal with problem solving as if it were a technical term with sharp boundary lines. It is more of a commonsense term that roughly stakes out an area we want to understand. And to this end, Skinner's definition works well.

Rule following, as defined based on RFT and as described above, represents the core of verbal problem solving. In this behavior's simplest form, it may be questionable whether the concept "rule" is adequate. In the example with the key, let's say that I visualize an earlier occasion of unlocking the door and then perform an action based on what I recollect. This activity can contain arbitrarily applicable relational responding, in which case it is verbal. If so, it is not made up solely of contingency-shaped behavior, but it is still questionable whether we should call it "rule following." A rule should, per definition, specify a contingency between behavior and consequence, with both behavior and consequence being contacted verbally. But regardless of these possibly borderline cases, rule following, and especially tracking, describes what we mean when in everyday language we say that we are solving problems. We relate different things before us to each other, and even to private events, which are also "before us" because we experience them from the same perspective from which we observe everything else. We relate all of these events—those contacted directly as well as those contacted indirectly—by using comparative, causal, hierarchical, temporal, and perspective-taking frames. Then, based on the many different options made available to us in this way, we take our action. Apparent contact with different behaviors and different consequences is the fundamental condition for this skill. When through relational framing we manipulate variables to make the appearance of a solution more probable (to paraphrase Skinner), we can use this skill to formulate self-rules that can in turn come to influence further behavior.

Some problem solving is strategic: What we want to achieve is clear; it is only the path to the goal that is unclear. A typical example is when someone gets lost on her way to a certain address. She finds herself in an unknown neighborhood and tries to reestablish her bearings by looking at a map and her surroundings, and by considering different possibilities. Another example could be what happens during an appendectomy if the surgeon realizes this particular patient's anatomy deviates from the usual, and thus new solutions are required. Yet another example is a psychologist working with a young boy who refuses to go to school, so that the boy can take up his schoolwork again. This one may border on being a situation where the goal itself is unclear. In this type of problem solving, rather than simply seeking the achievement of a known goal, we are faced with a variety of possible consequences, which necessitates comparing them with each other and making choices. A more typical example of this type of problem solving is a young person who has just finished college and is asking herself what to do with her life over the next few years. Other examples are choosing a spouse or partner, or an occupation. Yet other examples are what we often refer to as existential questions: "What do I want my life to stand for?" or "What is important to me?" Augmenting has a decisive function in this type of problem solving. Verbally constructed consequences that are globally desirable to the individual can come to control a wide range of behavior. I can relate specific actions and consequences that are close at hand to different values—to what I think is important in life.

In all of these situations, we relate both events we have experienced and those we have not to ourselves; to different possible behaviors and consequences. I can seemingly put myself in contact with everything from the beginning of the universe, humanity's purpose in the world, and dinner with my in-laws last night to the laundry I forgot to hang this morning, what I will be doing five years from now, my own death sometime in the future, and what will happen after that. I can relate any or all of this to something I am planning to do tomorrow and to how I feel about things in this very moment. And I can do all of these things while lying in my own bed. But most of the time I do this as a part of dealing with everything I actually encounter in life: when I talk to my coworkers, rebuild my summer cottage, bone up for an exam, have sex, organize political meetings, fix my car, or do my shopping in the mall. In all of these situations I encounter the world the way it is arranged. And in all of these situations, the ability to be both speaker and listener and to follow self-rules and thereby solve problems increases my flexibility in an almost unfathomable way. This activity—what in ordinary language we might call an ability to entertain ideas and solutions in connection with what we are faced with—is sometimes within RFT called pragmatic verbal analysis (S. C. Hayes, Barnes-Holmes, & Roche, 2001). This means that humans can abstract formal characteristics of events they are faced with and relate them to each other in a way that isn't necessarily circumscribed by these characteristics themselves.

SUMMARY

From the standpoint of behavior analytic assumptions, all human activity can be understood based on the contingency between a behavior and its antecedents and consequences. When a person has learned arbitrarily applicable relational responding, this has far-reaching consequences for how her learning activity continues. It allows antecedents to acquire stimulus functions that are restricted neither to the direct contingencies that the individual has actually encountered in her history nor by the physical qualities of different stimuli. Antecedents that have acquired their functions in this way, through relational framing, can now specify behaviors and consequences that are not yet present, and thus function as what we commonly call rules or instructions. This explains the human ability to act in relation to long-term consequences, rather than being completely controlled by direct contingencies. Or, in more everyday words, this allows us to put off immediate gratification.

As discussed, behavior that is influenced by verbal antecedents is called rule-governed behavior. Two different types of rule-governed behavior can be distinguished based on their historical reinforcing contingencies, and a third type interacts with both of these. Pliance is the fundamental type of rule-governed behavior; it involves rule following that helps us contact socially mediated consequences that are dependent on rule following as such. It is through pliance that we first learn to follow rules and instructions. Once this skill is learned, we can learn tracking: rule-governed behavior by which we contact consequences that depend on how the world is arranged, independently of the rule. The third type of rule-governed behavior, augmenting, is combined with pliance and tracking and affects the degree to which different consequences function as reinforcing or punishing. Augmentals function as verbal establishing operations. Figuratively speaking, you might describe these three forms of rule following like this: In pliance you seek what the rule giver "holds in her hand." In tracking you seek "whatever is on the map." In augmenting you seek consequences based on the value you assign to them.

Rule following vastly increases our ability to act flexibly in our social environment, as well as our physical environment. This ability seems to be the most general effect of verbal behavior (Catania, 2007). There is a cost to all of this, however: Rule following has certain side effects. In the next chapter, I will describe this dark side of our human ability to frame events relationally.

CHAPTER 7

The Dark Side of Human Languaging

It is easy to see the huge advantage human beings have in being able to put stimuli into arbitrary relations, especially when a set repertoire of rule-governed behavior is in place. This allows us to sidestep immediate gratification and deal with events before they take place. We can carry out long behavioral sequences, and we can act on consequences that are distant in time or in space or that are very abstract. At the same time, this force also has a dark side (Törneke, Luciano, & Valdivia-Salas, 2008).

RULE-GOVERNED BEHAVIOR AND INSENSITIVITY TO IMMEDIATE CONSEQUENCES

The most obvious and documented problematic effect of rule-governed behavior is an increased insensitivity to immediate connections between stimuli—to direct contingencies. Rules alter stimulus functions and override direct contingencies. This is a key advantage of rule-governed behavior, giving us the ability to sidestep immediate gratification, but it is not always advantageous. A number of studies illustrate this phenomenon (S. C. Hayes, Brownstein, Zettle, Rosenfarb, & Korn, 1986; Matthews, Shimoff, Catania, & Sagvolden, 1977).

Typically, the experiments are carried out as follows: The participants are given a simple task, such as pressing a button when certain lights come on. Some of the participants are told what they need to do in order to earn points (for example, "Press the button only when this specific light is lit"). Another group is given more general directions that do not specify the way in which points are earned. All of the participants are given immediate feedback when successful; that is, they can see when they have earned points. The advantage in knowing the rule is obvious, and those participants who know it initially earn their points faster than those in the control group. But after a while, the control group starts earning the same amount of points as the other group. They learn by trial and error. When both groups have started to earn points equally, the contingencies are altered so that all participants must use a new method of pressing the button to earn points. This change is made without any of the participants knowing the new way to respond to earn points. The various studies all demonstrate the same phenomenon: The participants who initially learned how to earn points by means of a rule have greater difficulty in discriminating the new contingencies. At this point, the control group more quickly learns how to earn points based on what has become reinforcing. The rule that was helpful in the beginning becomes an obstacle. It seems to stand in the way of quickly adapting to the altered and nonarbitrary relations between different events.

Most of us are acquainted with this from our own experiences. Our notions (rules) of how something ought to be or how something should be done can get in the way when we attempt to learn new things. We continue to do things that do not work, following rules that say they should work. We go on arguing for a certain position, even though it does not take us in the direction we meant it to, following rules that say we are right. We struggle to forget things we cannot forget, following rules that say it isn't good to think about these unpleasant things.

This brings us to the main subject of this chapter: how rule-governed behavior—as seen from an RFT perspective—contributes to what we usually call psychopathology. But before we take a closer look at how the different types of rule-governed behavior interact to create difficulties, let's take a look at an even more fundamental consequence of relational learning: that arbitrarily applicable relational responding necessarily means that we humans have a broad interface with pain.

A BROADENED INTERFACE WITH PAIN

Some things we like; others we do not. Some things taste good; others do not. Some things seem intriguing; others seem frightening. In the language of learning theory, we say that, for a given individual in a given context, some things are appetitive while other things are aversive. A common way new situations acquire either of these functions is through generalization. Things seem pleasant or unpleasant based on their resemblance to things we have encountered earlier, or because, through respondent learning, they have been associated with something pleasant or unpleasant. However, by means of relational learning, things can be related to each other independent of these nonarbitrary relations. Contextual cues can establish new arbitrary relations. This means that things that we have not yet encountered or that lack physical links with the things we have come across can nevertheless have functions for us. These functions can be appetitive as well as aversive. This is due to our ability to frame events relationally at the fundamental level of mutual entailment and combinatorial mutual entailment, as described in chapter 4.

Let's compare this with a situation where a nonverbal creature flees from a predator to find shelter. In such a situation, the creature can learn that a certain behavior leads to safety in the presence of a certain predator. Once the animal is sheltered, its position will have stimulus functions immediately connected with this place of shelter, without these functions being related to the predator. At this point, the function this place has for the animal is what we humans would call safety or security. In order for the function to be related to the predator, the events of the learning situation would need to occur in reversed order. According to the principles of operant and respondent conditioning, only if the animal's encounter with the shelter precedes its encounter with the predator can the shelter have stimulus functions related to the predator.

Suppose a human being were in a similar situation. Humans, too, would find functions of safety and security connected with the place of refuge. This spot will, however, also be related to the predator through mutual entailment. (In fact, the previous example, wherein an animal experienced the human phenomena "safety" and "security," is, strictly speaking, incorrect.) As humans, we cannot think of security without it being related to its opposite. In part, security is what it is to us precisely because of its not being its opposite. The two are related verbally, through mutual entailment. But for the animal in the example above, this place of shelter—as far as we know—is simply "this place": the place in and of itself, in the situation in which the animal finds itself at that moment. This is never the case for a verbally competent human being. Things are always related to their opposites, as well as to a number of other things. You might say that the ability to frame relationally means that a phenomenon that is very similar to generalization puts humans in touch with an almost endless number of events or stimuli, but without the type of restrictions that apply for generalization.

This has a decisive impact on the way we humans experience our reality. One important consequence is the functions our private events acquire. These private events are almost constantly present. Furthermore, all of our painful experiences contain such private events—things noticed and known only by the individual who experiences them. Through mutual entailment, private events acquire stimulus functions via their relations with the events the individual has experienced. If E acquires discriminative functions for D, then D, through mutual entailment, can acquire similar functions for E, as noted in connection with the laboratory experiments in chapter 4. Moreover, our ability to frame events relationally does not stop with these basic functions. Derived relational responding offers almost infinite possibilities. Starting with a specific experience, we can, for instance, put this into a comparative and temporal relation and thus contact "something worse after this." Private events become potentially painful in themselves, and through our ability to frame relationally, our interface with pain increases exponentially.

Imagine the following situation: You are spending time by the Mediterranean. The night is pleasantly warm. The day was wonderful—relaxing and invigorating at the same time. You get together with some good friends. The plan is to enjoy a nice meal together. People are talking all around you, and you hear bits and pieces of an intriguing story. The ocean is as smooth as a mirror just below the large patio where you are sitting. The waiters are beginning to bring your plates. Everything is just delightful; you are having a great time. Then there is the thought "If only Peter could have been around for this."

Even our good experiences are related to bad ones. Pain can be present anywhere. Indeed, we can be transported anywhere at the speed of thought without even moving. While this does provide a huge window of possibilities for us humans, it also creates a broader interface with pain. As long as pain is connected with a certain situation in the external world, we generally have the option of leaving to escape. There may be times when this option is not available, but quite often it is. However, where can a person run in order to escape from the pain he contacts through his own languaging? In the long run, the conditions under which we live as humans mean that there is no escape. Once the ability to put things in arbitrary relations is established, we are inevitably stuck—for good and for bad.

Consequently, pain is built into things that, in themselves, are far from painful. Consider a six-year-old girl who drew a picture at school. Her father says, "That's nice. You really did well on that!" Given her verbal competence, the child will also be in contact with the opposite of what is said, that is, "not so nice" and "not doing very well." She may not contact this at the very moment when her father speaks, but she may do so at a later time, such as when she tries to draw another picture and it doesn't turn out the way she wanted it to. She will not need anyone else to make the judgment that this is "not so nice" or that she "did not do well." Her ability to frame relationally puts her in contact with this in a specific context. There is no way out of this human predicament.

All of this is a part of the power of thinking. Thinking can take us anywhere, including where we do not want to go. Private events can become painful in themselves. Thoughts of pain, even types of pain we have never experienced, acquire stimulus functions from real pain.

Yet neither these thoughts nor other private events are the real problem (Luciano, Rodríguez Valverde, & Gutiérrez Martinez, 2004). They are a natural part of human language. Due to our ability to frame relationally, our interface with both aversive and appetitive functions increases exponentially. These are the conditions under which verbally competent human beings live. The real problem arises when, based on these private events, we begin to take actions that are followed by consequences and that don't work well. As you'll recall from the discussion of self-as-story in chapter 5, verbal behavior, especially of the private kind, can occur simultaneously with but independent of overt behavior. We can think, feel, and remember without this, in and of itself, having much effect on our environment. It is when we act overtly, and thereby contact consequences, that our lives are more generally affected. And it is when private events come to control overt action that the essential step in human psychological problems is taken, as seen from a behavior analytic perspective. This brings us back to rule-governed behavior. As discussed in chapter 6, rule-governed behavior can be influenced by private stimuli, such as self-rules. Many potential pitfalls arise when this human repertoire, in itself a result of relational framing, is combined with our broadened interface with pain. We all too readily learn to use this repertoire (rule-governed behavior) in a dysfunctional way in relation to thoughts, feelings, memories, and bodily sensations, especially in order to control painful aspects of these private events.

Even if this isn't a completely inevitable process, it is easy to understand why it takes place. Rule-governed behavior is successful for humans in many different areas, and particularly in the area of avoiding danger and other events that could have negative consequences in our lives. It is usually helpful to clear away obstacles to what we want to achieve. When thoughts, memories, and feelings acquire stimulus functions from events that have been or could have been aversive, it is natural for us to deal with these phenomena in the same way we are used to dealing with other things that hurt us or impede us. Simply put: Make a plan for getting rid of the difficulty or obstacle and execute the plan. It is in connection with this strategy—intentionally trying to control private events—that we most clearly can see the dark side of relational framing. Especially in light of the fact that rule-governed behavior has a tendency to continue even in situations where it does not fulfill its purpose, as evidenced by the experiments described earlier in this chapter. This behavior is insensitive to direct contingencies. So if we try to control painful private events by following rules saying that these phenomena should or must be eliminated, the fact that these phenomena are not subject to our efforts might not stop us from continuing to follow such rules. As a result, we risk ending up in vicious circles that restrict our lives.

PSYCHOLOGICAL PROBLEMS AS A CONSEQUENCE OF RULE-GOVERNED BEHAVIOR

For the sake of clarity, I will go through the problems connected with each type of rule-governed behavior separately. In real life, however, they are all continuously interacting with each other. The most essential issue here is how augmenting interacts with the two more basic types of rule-governed behavior to alter the function of present contingencies, creating different types of pitfalls.

Problems Connected with Pliance

The fundamental problem with pliance is its insensitivity to direct contingencies. A person who constantly seeks reinforcement and attempts to avoid punishment by trying to please others will have problems contacting other consequences, appetitive as well as aversive. Using the imagery from the end of the previous chapter, we might say that a person who primarily seeks things that "the rule giver has in his hand" is going to risk missing a lot of other things that "are on the map."

Here's an example: A person who visits art museums primarily because this is what an intellectual person should do, or in order to gain the approval of someone particular, lowers his chances of enjoying the art or learning more about art as such. Parts of his potential experience will be blocked because focusing on the approval of the other, for example, is incompatible with focusing on the details of what is displayed at the museum. The consequences that the person would have encountered if he had engaged his interest in the artwork for its own sake simply are not contacted, as this behavior never occurs. A major difficulty when pliance dominates a person's behavioral repertoire is that consequences that are dependent on other people are not as predictable or controllable as other types of consequences. Someone can, for example, love mainly in order to be loved back, but this is not necessarily what results. A person might listen to others primarily to have them listen to him, but this does not always happen either. If someone repeatedly reaches out to another person mainly because he hopes to get that person's approval, he will probably be disappointed, off and on. The most obvious consequence for people exhibiting generalized pliance is that they become vulnerable to social whim.

This applies to what might be called counterpliance, as well, such as when a person opposes an authority just because "You should never follow orders." Even when these actions backfire on the person in the long run, he tends to resist or act in opposition when an authority tells him to do something. It is as though he cannot comply with authorities—even in situations where he would obtain reinforcing consequences by doing so. Although in this case the person is acting in opposition, he is following a rule controlled by the same functional links as when obeying. This means he does what he does in order to achieve socially mediated consequences based on what the rule lays down—only the other way around. Functionally, counterpliance is a type of pliance, and it can become a problem if it dominates the behavioral repertoire.

When pliance is generalized and dominates a person's actions, two key problems arise: First, this may stop him from learning successful tracking (S. C. Hayes et al., 1998). As tracking is a skill that increases behavioral flexibility enormously, this can be problematic. And second, the person remains very sensitive to the consequences offered by others. There is a risk that the only governing consequences will be those managed by others, and this will make the individual very vulnerable in relation to his social environment. Over time, consequences become reinforcing or punishing simply because they are controlled by others.

If we follow the rule, explicit or implicit, "Comply with the wishes of others so that they are pleased with you" as the essential rule in life, we run a high risk of placing our lives in the hands of others and straying from our own goals and desires. Whether this becomes a problem depends, in large part, on how successful we are in getting the approval of others—and always getting the approval of others is not easy. In most human settings, if we follow this rule, we run a high risk of not getting enough of the approval we seek. And if we don't get what we want and the rule is still in control, we will make even greater efforts to receive approval. Even then, we still might not get enough, so we try even harder and end up in a vicious circle. This can lead to a very limited life, a life that lacks contact with a number of other potential reinforcing consequences. These potential consequences—which cannot be contacted without tracking—are blocked. This is a classical clinical problem that has been described at least since the days of Freud. In therapy sessions, you may notice generalized pliance when a client seems to tune in to "what the therapist wants" in a rigid way. Counterpliance in relation to the therapist is another form of the same phenomenon. One example is a client who never adheres to instructions from the therapist, from a stance of "No one is going to tell me what to do."

To understand how this form of rule-governed behavior turns into a more comprehensive problem in a person's life, we also need to understand the interaction between generalized pliance and augmenting. Augmenting may contribute to the persistence of generalized pliance, as well as to its increase. This occurs when this problematic type of pliance is controlled by abstract and verbally constructed consequences connected to the primary, socially mediated consequences. Most people whose behavioral repertoire is dominated by problematic pliance are not behaving this way simply based on immediately available consequences, such as getting someone's approval in the here and now. More all-embracing premises about what is important in life usually come into play. I will return to this issue below, in the section "Problems Connected with Augmenting."

Problems Connected with Tracking

The simplest form of tracking consists of that which specifies a direct contingency, that is, a behavior and a consequence, where this consequence will follow such behavior independently of the rule as such; for example, "Take a painkiller, and it will hurt less." The behavior is performed, and the consequence is contacted. The capacity for tracking is strengthened in this way, and the person's behavioral repertoire is expanded. The probability increases that the person will follow new and possibly more complicated tracks. Tracking can also be used in the pursuit of long-term consequences, such as getting good grades at the end of the semester. This augments the reinforcing function of direct, short-term consequences for different types of actions, such as handing in school assignments on time, being in class for every lesson, and so on. This is an example of how augmenting interacts with tracking.

However, tracking can become problematic in several ways. One example is when tracking is used in contexts where it cannot work. For instance, take the rule "I have to be unconstrained," which of course is impossible to follow. A naturally spontaneous behavior cannot, by definition, be under the command of instructions. A person who acts based on the rule "I have to be unconstrained" will not, as a result, have the experience of being unconstrained. Tracking cannot function when the behavior specified in the rule is incompatible with rule following as such. It would be like trying to follow the rule "You must not follow rules."

Tracking also becomes problematic when the track does not correspond to the way the world is arranged and rule following therefore does not lead to the specified consequence. Let's use the rule "If you stop smoking, you will feel better" to illustrate this. Most people who stop smoking won't primarily contact that kind of consequence, at least not initially. An important point is that if you follow incorrect rules such as this one, the target behavior will probably wane, since the specified consequences that were to sustain the rule following do not occur. If someone stops smoking specifically to "feel better," that person is likely to take up smoking again in a relatively short time.

Yet continuing in actions based on incorrect rules can potentially become troublesome in a different way. Take the rule "In order to recover, I have to avoid pain." Based on this rule, the individual might act in a way meant to keep pain away, and as a result, he won't experience that the short-term and long-term consequences specified in the rule (not feeling pain and recovering) are not coordinated over time. The person takes action to avoid pain and pain is avoided, so the behavior is negatively reinforced. In addition, the connections established in the rule, between recovery and not feeling pain, create a consequence of "doing things the right way," which also sustains this type of tracking. In this example, there is a problem with the short-term consequences specified in the rule. It's true that you might not experience pain if you don't perform a certain physical activity. But the way this world is arranged, pain might play a part in the process of recovery. If the person acts in accordance with a rule stating that pain is incompatible with recovery, then tracking of relatively short-term consequences (limited physical activity → no pain) will block tracking of more long-term consequences (activity → pain → recovery). This becomes a vicious circle where well-functioning short-term tracking does not work in the long run. However, the fact that this tracking works in the short term means there is a risk that the governing function of the rule as a whole is reinforced. This is because the rule coordinates the short-term consequence of having no pain, a consequence that is actually contacted, with the long-term consequence of recovery. On top of this, there is the reinforcing consequence "doing things the right way."

This paradoxical effect is a fundamental process in many common clinical problems, including depression, anxiety disorders, and different types of addiction. An example would be a person with social anxiety disorder. When the person acts according to the rule "I cannot leave home because it will give me anxiety," the person is often correctly tracking the short-term consequence of not feeling as anxious by staying at home. But if this behavior continues over time, this kind of tracking may have the opposite effect, causing social anxiety to increase and possibly leading to depression. Another example would be a person who is addicted to alcohol. When he tracks the rule "I will feel better if I do some drinking now," it may work in the short term. However, if he continues drinking, his problem will not disappear, and his experience of not being able to control his life in the way he wants to will probably increase; and at the same time, he will not track the long-term and destructive consequences of continuing to drink. This person is stuck in a limited, and in this case insufficient, behavioral repertoire when it comes to tracking long-term consequences.

These examples may seem to imply that the person doesn't understand the long-term consequences. In fact, this is rarely the case. People who withdraw in order to feel better and avoid anxiety usually know that this increases anxiety in the long run. People who drink as a means of coping usually know that drinking will worsen their problems in the long run. Like any other form of private verbal behavior, the private verbal behavior of "understanding" can occur without having any important effect on overt behavior. The problem is not primarily a lack of understanding; it's a result of following a rule and contacting the short-term consequences. This important point bears repeating: It is not primarily a matter of understanding; the crucial element here is rule-governed behavior followed by consequences.

Negative reinforcement clearly plays a role in the examples above. By not contacting social settings, you can avoid the experience of social anxiety. By drinking alcohol, you can decrease anxiety and uneasiness in the short term. But other reinforcement contingencies also come into play. Augmenting plays an important role in sustaining poorly functioning tracking, just as it does in connection with problematic pliance. As mentioned above, "doing the right thing" alters the functions of ineffective tracking so that the individual continues to follow the rule in spite of the long-term consequences this leads to. We humans easily get stuck in following rules that do not work well, especially rules that govern behaviors for controlling fear, downheartedness, shame, grief, or feelings of loneliness. We do this in an attempt to reach goals along the path toward something "further still." The essential function of this direction—this "further still"—is connected with problematic augmenting.

Problems Connected with Augmenting

There is a common denominator between augmenting and pliance: Both are sensitive to social whim. In the case of pliance, this is a result of the consequences specified by the rule being arranged by the social community. The verbal networks that make augmenting possible are also arranged by the social community, though in a more complex way; they are shaped by the social interplay that lies at the very base of languaging. The relational networks that serve as augmentals, intensifying the governing function of consequences, can be very abstract. And the more abstract the governing consequences, the harder it is for immediate consequences to have an impact on behavior. The fundamental problem in connection with augmenting is that its interactions with pliance and tracking lead to increased insensitivity to direct contingencies.

Let's return to the problems connected with pliance. As mentioned, augmenting can easily make the problems with pliance persist and increase. Although there are individuals who seek appreciation from others in order to obtain tangible forms of reinforcement, generalized pliance usually occurs under the influence of more abstract, verbally constructed consequences. Abstract consequences like "being a good person," "being successful," or "being lovable" may have become established as the ultimately desirable consequence for a particular individual. These types of consequences may be coordinated with specific actions, like doing what others do, agreeing with what others say, staying within a certain pattern of behavior, or never saying no. This type of learning history strengthens an arbitrary relation of coordination between, for example, "being lovable" and adapting to others in various ways. This will, in turn, affect the stimulus functions of a behavior that involves not adapting. For a person like this, the various forms of "not adapting" will have punishing functions. This is because "not adapting" is in opposition to "being lovable," which is the self-evident aim or value for this individual. Let's take a detailed look at what is going on in problematic augmenting based on generalized pliance when a person systematically acts on a self-rule such as "To achieve what I want, I need to act in a way that makes me feel good, and I feel good when no one criticizes me and when I get appreciation from people around me."[1] This is the result of advanced relational framing:

1. Certain thoughts and feelings that are evoked by people's criticism or their failure to express appreciation (for example, "I am bad; I'm not the way I should be") are put in opposition to actions that are aimed at specific goals. In essence, "If I have these thoughts and feelings, they will stop me from acting in a certain direction."

1 Note that the person does not necessarily observe any thoughts when the rule following occurs. He may report having these thoughts or he may not. The rule may be implicit. What is essential is that the person is in fact acting in this way.

2. The person's goals are put in coordination with feeling good; that is, "I need to feel good to be able to act toward these goals."
3. Feeling good is established as a necessary component (causal or conditional framing) of reaching the things that really matter to that person.

Another example of problematic augmenting is a depressed person who dwells on past events. If asked whether going over things again and again seems to lead him anywhere, he will probably acknowledge that it does not. Still he keeps doing it. Negative reinforcement plays a significant role here. This activity helps him avoid something he experiences as painful. Perhaps he avoids other thoughts by dwelling on the past. But negative reinforcement is not the only governing consequence. This person's rumination probably belongs to a functional class of behavior that we would call problem solving. This is what you are "supposed to do" to deal with problems. So this is actually another example of how a behavior is reinforced by "doing what I need to do in order to feel good." Thinking things through is seen as the right thing to do. We might ask why this type of problem solving becomes so intense for a certain person in a certain situation. People whose thoughts constantly refer to the past often say that this is an attempt to feel better. And why is feeling better so important? Because it is seen as a means of reaching other important goals or values in life. This rule—"I have to work to get rid of my depressed feelings in order to feel good, and I need to feel good to be able to achieve what is important in my life"—functions as an augmental that alters stimulus functions in various ways. The thoughts the person is avoiding by dwelling on past events become even more aversive through augmenting, since they are no longer just painful in themselves; they are now in opposition to reaching important things in life, as well. At this point, the augmental intensifies efforts to get rid of the undesired thoughts because the rule specifies that this is what it takes to achieve important goals and live in accordance with one's values.

Imagine a person who is experiencing hallucinations. Let's assume that this person is tracking the rule "If I stay in bed today, I will not hear voices." When asked why it is important not to hear voices, the person might respond that life is easier to handle without hearing voices, that he doesn't risk being locked up in a mental institution, or that he wants to get away from thoughts of going insane. This tracking may work in the short term, but it does not track more long-term consequences. As a result, the behavior becomes part a long-term vicious circle, as described above. Augmenting could potentially complicate this situation further through a self-rule like "I have to make sure I don't hear voices, since a normal person doesn't hear them." As a result, experiencing hallucinations becomes even more aversive, since they've been put in opposition to being a normal person. This further motivates the behavior that, according to the rule, is supposed to control the hallucinations. The process as a whole now goes beyond the original painful experiences—experiences that can be difficult enough in themselves. It is now also a question of "being a normal person."

The same principle can be seen in processes involved in dependent personality disorder. A person with this diagnosis will often behave in ways aimed at avoiding feeling a lack of support, whether by openly appealing to others or by being passive. Let's assume that a rule governing this person's behavior is "In order to be able to do what I want, I need to focus on feeling supported." This rule intensifies the aversive quality in the experience of not feeling supported, since that experience now stands not only for the actual uneasiness, but also for obstructing the life the person desires. This increases the motivation to act in a way that is supposed to bring about feelings of support or avoid the opposite. And as the experience of support naturally varies in a social environment, this can all too easily lead to a vicious circle wherein the more this experience eludes the person, the more he finds it important to achieve.

All of the clinical examples above illustrate the same foundational course of events and the same human dilemma, regardless of diagnosis. Rule-governed behavior, in the shape of pliance and tracking, allows us to act to achieve more long-term consequences. However, it also necessitates a degree of rigidity. This rigidity can increase under the influence of augmenting, especially when the consequences sought are either very abstract or distant in time and space.

Let's return to the definition of augmentals. They are rules that come close to what we in everyday language would call assumptions, premises, or starting points. Augmentals are often implicit and seem matter of course. We just take them for granted, and that is precisely how they influence our behavior. They are verbally constructed, complex relational networks that govern our behavior by operating on the degree to which different consequences function as reinforcing or punishing. They function as a type of verbal establishing operations. Augmenting is regarded as the most advanced form of rule-governed behavior (S. C. Hayes et al., 1998). Just to be clear, all rule-governed behaviors have good and bad sides, but in this context I am emphasizing their problematic effects.

Note, however, that from a behavior analytic perspective these assumptions or starting points, in and of themselves, are not the problem. The problem is not the rules. The rules are simply words expressed by others or private events discriminated by the individual, or, in many cases, they are neither expressed by others nor discriminated as a private event; they are implicit. When put into words, they are verbal abstractions of actions, as discussed in chapter 6. The real problem isn't the rules; it's rule *following*. When actions result in consequences that turn into life problems, these problems arise, in part, because certain actions simply do not give us what we want, and, in part, because these very actions preclude other actions that could have been more likely to lead to reinforcing consequences.

RISKS RELATED TO THE DOMINANCE OF INDIRECT STIMULUS FUNCTIONS

Once a repertoire of relational framing is established in a given individual, derived or verbal stimulus functions can dominate over directly established functions. This was shown in the experiments mentioned in the beginning of this chapter, where a rule continued to govern participants' behavior even when the contingencies were altered in a way such that the rule no longer corresponded to how the environment was arranged. In all likelihood, this occurs because our experiences of acting on indirect stimulus functions have been highly reinforcing in so many areas. Simply put, language works.

When someone says something, they utter a sequence of sounds. This becomes very clear when we listen to someone speaking a language we do not understand. In that case, we only contact the direct stimulus functions: the character of the sounds, perhaps the rhythm and speed of the utterances, and so on. If we've previously heard that language spoken, it may provoke memories from those times or situations. Nevertheless, our experience is still mostly controlled by direct stimulus functions—the immediate connections between these sounds and our direct experiences. This tends to diminish when we hear a language that we've mastered. The indirect stimulus functions come to the forefront. We understand what the person is saying and act based on this. Yet all we truly hear, even in this case, is combinations of sounds. However, because both speaker and listener have the same way of putting particular sounds in relation to each other, and in relation to a number of other things, we understand what we hear. These relations are, in themselves, arbitrary. We are playing the same social game together.

One way of comprehending the natural flow by which indirect stimulus functions dominate is to try to contact direct stimulus functions in a context where this does not normally pay off. As a reader, you can try this now, by looking at the text you are reading and focusing on the contrast between the paper's color and the pattern made by the printing ink without making a note of what the text means. You may find it remarkably difficult; the meanings of words are so evident to us. This does not necessarily indicate that it is impossible to switch this meaning off, or that the indirect stimulus functions must dominate. It might be possible that, perhaps by means of some meditation or focusing technique, you can unlink the meaning from the actual text, so that you simply note the visual contrast between black and white. However, this is not the fundamental attitude we have learned to take in relation to written text.

Another way of experimenting with direct and indirect stimulus functions is to say a word aloud—quickly and repeatedly—for a while. If you haven't done this before, give it a try now. Take the name of the city or town where you were born and repeat it quickly and loudly for a while.

What did you experience? When doing this, most people experience short moments when the meaning disappears and they only hear a sound. How does this work? Based on RFT you might say that the context that normally sustains indirect stimulus functions is altered, and this is what makes the functions diminish. Performing this exercise changes the aspect of the context referred to as C_{func} in RFT. Indirect stimulus functions are not inherent; they arise as a result of how we behave, sometimes together and sometimes by ourselves. It is possible to alter these functions by influencing contextual factors. In the examples above, the effect is obviously transitory. The normal context is solidly present.

Indirect stimulus functions do not dominate everywhere and in all circumstances. Humans live in a world where both direct and indirect stimulus functions influence our behavior. In some situations, direct stimulus functions dominate. If you are a skilled pianist and you are playing one of your favorite pieces by heart, then you are doing this mainly under the influence of direct stimulus functions of the moment. If someone were to start speaking to you—say they suggest that you play it in a different way—you would notice the disturbance, but you could undoubtedly let this go and really immerse yourself in playing this music so that you would still mainly be under the influence of direct stimulus functions. To continue with a behavior and simply let your next step be influenced by whatever you face directly is an ability that is still intact in humans, even if it must compete with rule-governed behavior.

The dominance of indirect stimulus functions need not be a problem. Most of the time, arbitrarily applicable relational responding increases the flexibility of our actions. That it increases the risk of rigidity in other situations is simply the other side of the same coin. Yet the risk of being overly influenced by a strict dominance of indirect stimulus functions is real. Earlier in this chapter I described this in connection with problems related to the different forms of rule following. When this dominance becomes strong, RFT uses the term *fusion* to indicate that certain actions are completely dominated by, or fused with, indirect stimulus functions. *Fusion occurs when certain verbal (indirect) stimulus functions dominate over other potentially available stimulus functions, both direct and indirect* (Strosahl, Hayes, Wilson, & Gifford, 2004; S. C. Hayes, Strosahl, Bunting, Twohig, & Wilson, 2004). I can act wholeheartedly based on the assumption that licorice tastes terrible without ever having tasted it, regardless of other potentially available direct stimulus functions (how it would taste if I tried it) or indirect stimulus functions (other opinions about licorice). I can act completely in line with the assumption "I am not able to talk to him," even though other direct stimulus functions (what would happen if I talked to him) or indirect stimulus functions (other views on what I am able to do) are potentially available. Again, it is important to emphasize that it is fusion *in action* that can become problematic. You might put it this way: Fusion *is* action; it does not refer to a mental process that in some way precedes the behavior. The potential pitfall connected to dominance of verbal functions is fused behavior followed by consequences, not inflexible thoughts in themselves.

EXPERIENTIAL AVOIDANCE AS A CENTRAL PROCESS IN CONNECTION WITH PSYCHOPATHOLOGY

A fundamental point made in this chapter is that the dark side of arbitrarily applicable relational responding becomes most evident when rule-governed behavior is focused on controlling private events. The reason for this behavior is that private events can easily attain aversive functions through arbitrarily applicable relational responding. This was also a prominent theme in the preceding survey of how different types of rule-governed behavior can turn into behavioral traps. We all too easily come to focus our actions on controlling private events. This type of rule-governed behavior, called *experiential avoidance,* is defined as actions aimed at controlling or eliminating affects, thoughts, memories, and bodily sensations (S. C. Hayes, Wilson, Gifford, Follette, & Strosahl, 1996). The outcome of this behavior is generally paradoxical. It may work in the short run, but in the long run it heightens our risk of actually increasing the type of experiences we are working to avoid. At the same time, life as a whole becomes restricted and life satisfaction wanes. I have illustrated all of this above, in the passages on the negative effects of rule-governed behavior aimed at controlling private events, and extensive scientific data confirm that experiential avoidance can be central to psychopathology (Karekla, Forsyth, & Kelly, 2004; Chapman, Gratz, & Brown, 2006; Campbell-Sills, Barlow, Brown, & Hofmann, 2006; Chawla & Ostafin, 2007).

The origins of this problematic behavior are straightforward. Threatening situations have aversive stimulus functions and become the object of avoidance for all organisms. For us humans, private events readily acquire aversive stimulus functions through relational framing, and thereby also become objects of avoidance. At the same time, our interface with these events is immense, if not limitless. As rule-governed behavior is established in the child, a number of advanced behaviors become available and may be turned toward controlling thoughts, feelings, and bodily sensations. If rules then are established that entail a relation of opposition between private phenomena that are felt to be aversive versus a satisfactory life, these rules are likely to function as augmentals. These augmentals can then establish the private events in question as even more aversive, so the importance of avoiding them becomes even greater, further increasing the motivation to deliberately focus actions on controlling these phenomena. In this way, experiential avoidance is established.

As the ability to frame relationally gradually becomes more complex, an increasing number of events become related to other stimuli in a wide variety of ways. More feelings and thoughts can therefore be coordinated with that which is aversive. These feelings and thoughts, in turn, can be put in opposition to that which is desired, and thereby also become the objects of various avoidance strategies. *The problem with experiential avoidance is that the human ability to control private events is quite limited.* To a great degree, these phenomena are out of our volitional control. Thoughts emerge, and feelings, memories, and bodily sensations are activated in an automatic way in countless situations in life, making it impossible for rules that propose control of private events to work effectively. Another problem is that the more you try to follow rules like these, the greater the risk that in the longer term you'll wind up more closely in touch with what you have been trying to avoid (Wegner & Gold, 1995; Abramowitz, Tolin, & Street, 2001). Yet another problem, and the most important one, is that when an individual commits himself to this form of deliberate avoidance, other behaviors, which might have led to positive reinforcement, are precluded. The long-term result is an increased risk of living in a way that is ultimately unsatisfactory to the individual.

TO STRUGGLE WITH ONESELF

A human's experience of himself, and his ability to verbally discriminate himself and his actions, are a prerequisite for more advanced rule-governed behavior. Other species can also discriminate their own actions, but because of our capacity for relational framing, the complexity of human self-discrimination is enormously greater (D. Barnes-Holmes et al., 2001). We humans can put our own actions into a number of different relational frames. This means we can contact "this action—earlier or later on" (temporal framing), "not performing this action" (oppositional framing), "more of this action" (comparative framing), or "this action over there" (perspective framing), for example—and all without the action having to be at hand or even having occurred previously.

In addition to this, we are able to put our experience of self in relational frames. Through verbal behavior, we can learn to discriminate that "*I* am the one who is performing this action." We do this through verbal perspective taking, as described in chapter 5. This skill is combined with rule-governed behavior to form self-generated rules where the individual is both speaker and listener.

The ability to discriminate oneself in this way can be affected by the problematic complication discussed above: the dark side of relational framing. If thoughts can acquire aversive functions from essentially any direction through relational framing, then my thoughts about myself can also acquire aversive functions. This can take place through direct experience. For example, when I experience something painful, this is directly related to my experience of myself. It can also take place through other people's narrative of myself, such as if I am often told that I'm not the way I ought to be, that I do things the wrong way, that I'm bad, or that I'm stupid. If direct and indirect means are combined, the probability increases that I will develop a narrative of myself that is painful in different ways, and through this narrative I risk having my life circumscribed. However, even if I do not have powerful experiences like these—either of pain in itself or of frequently being presented with an aversive story of myself—I still will not escape this possibility. Although my experiences and the story of myself that others have handed to me might be dominated by appetitive functions, my verbal ability will necessarily also put me in contact with aversive functions in relation to myself. All organisms constantly discriminate between aversive and appetitive stimuli. These discriminative responses increase in a dramatic way in connection with the ability to frame events relationally, especially when framing comparatively. Since humans, through verbal perspective taking, learn to relate to themselves as objects, the following question becomes inevitable: Am I appetitive or aversive? In more everyday language, the question might be phrased along these lines: Am I good or bad? Am I the way I should be? Am I good enough?

Just as both direct and indirect aversive stimulus functions are related to private events in general, they are also related to the experience of "me."

Because my private events are in a very special and direct relation with my sense of self, it is easy to understand that they will acquire functions that are central to my behavior. They do, after all, take place within me. In the discussion of how our experience of self originates, in chapter 5, I described how our stories of ourselves are connected with risks of rigidity and of falling into verbal pitfalls. The more I act in fusion with my story of myself, the greater my risk of acting in a way that in the long run has a negative and restricting effect on my life. This is particularly applicable if my direct and indirect learning history has resulted in a limiting story that increases the probability of problematic rule following.

Let's suppose that my story of myself describes me as someone who "can suddenly do something completely crazy" or who "cannot make it alone." It isn't hard to see how a narrative like this—if I act in fusion with it—promotes a kind of rule following that sustains guardedness and behavior that lacks independence. Such a narrative also assigns indirect aversive functions to aspects of self-as-process. A certain feeling that I notice, say feeling insecure, can become an obstacle to action, an antecedent to avoidance. This type of avoidant behavior tends to result in negative reinforcement because, in the short run, my feeling of insecurity decreases when I act in a guarded way. Following this rule can therefore potentially increase my tendency to act in fusion with the story, which will come to further dominate my behavior as a result. And all of this can take place even in situations where this behavior causes severe losses in my life. Experiential avoidance has become a trap.

A number of psychological problems and psychiatric diagnoses can be analyzed in a corresponding way. Responsibility-focused stories of self are often important elements in connection with obsessive-compulsive syndromes. Stories dealing with guilt are often key in depressive conditions. Stories about the individual's own body and appearance have central functions in connection with eating disorders, and narratives about oneself and one's need for interpersonal affirmation often have these functions in connection with borderline personality disorder. These are just a few typical examples.

Many people explain that they seek psychological counseling because they feel low in self-confidence or self-esteem. Many of these individuals express their problem in terms such as "There's something wrong with me" or "I'm not normal," or they seem to assume that there is "something wrong inside" that causes their dilemma or their symptoms. Many of these descriptions of the problems are problematic in themselves and constitute some of the verbal traps that restrict the individual. Still, these phrases also point out that an important part of the human dilemma is connected with how we act in relation to our experience of self. This experience of self can acquire aversive functions through verbal learning and can be further complicated through the problematic types of rule following described in this chapter, which lead to experiential avoidance.

SUMMARY

Human language is enormously useful, yet it also has some problematic side effects. The behavioral repertoire of derived relational responding results in the human capacity to transform stimulus functions by way of mutual entailment and combinatorial mutual entailment. As a result, the possibilities of contacting nonpresent events increase dramatically, so aversive stimulus functions can potentially become omnipresent. In addition, our capacity for relational framing also makes rule-governed behavior possible. This too is very useful, but it opens the door to getting caught in different types of behavioral traps. Among the types of rule following that I have described, augmenting seems to be potentially the most problematic. When we follow augmental rules about what is necessary in order to have a good and meaningful life, many private events attain strong aversive functions. Deliberate efforts to control them then seem to offer the only way forward. However, this path all too often becomes a dead-end street. There's a certain futility in this kind of rule following, since these private events aren't typically subject to efforts to control them, and, paradoxically, the very effort to avoid these events actually increases the likelihood that they will occur. Compounding the problem, and perhaps worst of all, these efforts can turn into an essential aspect of life, resulting in other behaviors being abandoned—behaviors that might have increased the probability of contacting positive reinforcement in the long run. These vicious circles are established by the basic problem of acting in fusion with these rules. Because many of these rules are self-generated and also often specify what the person should do in regard to "himself," an important part of these problematic behaviors is enacted in relation to our experience of self.

It isn't inevitable that we get caught in traps like these, and even if we end up on such a dead-end street, it isn't necessarily a terminal point. Our behavior can be influenced by a number of different contextual factors. Therefore, changing contextual factors as a means of diminishing destructive rule following is the goal of any psychological treatment based on the principles presented in this book. It is my aim, in part 3 of the book, to explore how to approach this type of treatment.

PART 3

Clinical Implications

CHAPTER 8

Learning Theory and Psychological Therapies

All psychological therapies must deal with two separate arenas, two principally different sets of conditions. One arena is that shared by therapist and client—their time spent in sessions. The other arena is the rest of the client's life, where he encounters the problems or difficulties that have made him seek help. The first is the only arena the therapist can influence directly, as she is present there. At the same time, the second arena—the client's life outside of therapy—is, of course, in the long run more important to the client. That is where change needs to take place.

This provides the therapist with two possible ways of exerting influence, as seen from a behavior analytic perspective: through client-therapist interactions when the client's problematic behavior is present during sessions, or through the therapist's ability to influence the client's rule-governed behavior. In the first case, if the client behaves the same way during therapy sessions as he would have out there "in real life," there is a possibility that the therapist can offer different circumstances from those that normally influence the client's actions, giving the client an opportunity to learn something new. An example is a person who typically turns quiet and withdraws from others when he feels disappointed and sad, and for whom this behavior is a problem. If this occurs in his interaction with the therapist, the therapist may act in a way that creates circumstances that change the client's behavior. Perhaps the therapist notices how the client withdraws and calls his attention to this while also encouraging a different type of behavior. If the client tries a new way of behaving, like expressing his disappointment, the therapist can act in a way that reinforces this new behavior. This process is also a fundamental part of exposure therapy. For example, a person with obsessive-compulsive problems may not dare to keep knives within reach for fear of harming others. In this case, therapy will typically include creating situations, arranged by the therapist and client collaboratively, wherein knives are within reach. This may occur in the consulting room or in a place where this problem is usually triggered, such as the client's home. The therapist works to support new behavior, in this case for the client to have access to or handle a knife in a way he has earlier avoided, and she tries to arrange for consequences that reinforce this new behavior.

Both of these examples demonstrate how the therapist can build on behavioral principles to arrange a new context for the client's problematic behavior when it occurs, and in this way increase the probability of behavioral change. In the latter example, the therapist arranges a setup of antecedents and consequences for both problematic and new behavior. Of course, the therapist herself is also a part of the client's context when the two of them are interacting. However, given what I mentioned about therapy's two arenas, it is evident that changes that occur when the therapist is present are not the primary goal. It is change in the second arena, the client's own life, that we are aiming for in therapy. If this is to happen, generalization is necessary; the client learns something new as he works with the therapist, and this is subsequently generalized in his day-to-day life. But for such changes to be lasting, the consequences that face the client "out there," in real life, must reinforce this new behavior as well. If not, there is a risk that the newly learned approach from therapy won't be firmly established in the client's ordinary life and thus will be easily extinguished. The therapist does not control consequences that take place in the second arena, so in their work for change, therapist and client must take this into consideration as they create a treatment plan. In the client's day-to-day life, are there natural reinforcers for the client's new way of behaving? To go back to the example of exposure therapy, we can assume that in most cases there is more reinforcement for the ordinary use of knives than there is for always avoiding them. Likewise, expressing one's disappointment and sadness, rather than withdrawing, is probably reinforced, at least in certain contexts, though not necessarily in all.

Paths to desired changes can thus involve a process along these lines: The client practices new behavior in his interactions with the therapist and encounters new consequences that reinforce this behavior. He then acts in the same way in his day-to-day life and encounters even more reinforcing consequences, making it more likely that the new behavior will occur in other, similar situations. In collaboration with the client, the therapist has set up a new context consisting of both antecedents and consequences, and this leads to a change that is subsequently generalized. This path to change is central in connection with many problems that, at a cursory glance, may seem to differ considerably. To have problems with intimacy and with telling others what you think and feel may, for example, seem very different from an obsessive-compulsive symptom like avoidance of handling knives. Seen from a behavioral perspective, however, both the problems themselves and the paths to change may be similar in a fundamental way. This is, of course, because behavioral principles are not primarily about psychological problems or clinical diagnoses. Behavioral principles are about human behavior as a whole and are therefore universally applicable.

From a behavioral point of view, an important part of the work toward change is that the sessions in which therapist and client meet (the first arena) provide opportunities to learn new things in this direct way. But another element is also involved. The client tells the therapist about things that happen in his life when the therapist isn't present (the second arena). They discuss the client's behavior: what he has done and what he has experienced. This occurs naturally in all types of psychological therapy, regardless of the school of thought. In fact, it is regarded as such a natural fact that it is often seen as unproblematic, obvious, or natural. However, this isn't the case from a behavioral perspective. Instead, we question how this is possible. How can two people, sitting in a room, exchange series of sound combinations and gestures, and in so doing, contact things that have happened earlier—maybe years ago—or things that have not yet taken place? And how can talking in this way influence what happens later? Hopefully, parts 1 and 2 of this book have provided some answers to these questions, or at least a clearer idea of the processes involved. The possibilities of influencing events through languaging are a result of our learned capability of framing relationally, in the way described in part 2 of this book.

The area in which arbitrarily applicable relational responding has the greatest practical consequences is our ability to create and follow rules and thereby influence behavior that is not ongoing. This applies to the behavior of others, as well as our own behavior. Just as this is a natural part of all aspects of human life once we have learned the ability to engage in rule-governed behavior, it is also a natural part of all psychological therapy. It will be taking place regardless of whether the therapist emphasizes it.

Let's return to the examples above: learning to tell others what you think and feel, and getting over painful thoughts of what you could do if you were holding a knife in your hand. If the first client communicates his disappointment to the therapist, and their subsequent interaction then differs from what the client has experienced outside the consulting room, this means he can learn new ways of behaving. If the second client actually handles a knife in a successful way, and if his fear that he will do something he doesn't want to do isn't realized, then he too has learned something new. Most likely, even without the help of the therapist, the client will then be able to formulate rules along the lines of "If I act like this at home, things might turn out more the way I want them to," simply based on his experience in therapy.

Of course, this chain of events isn't inevitable, and it doesn't necessarily follow that formulating rules like this will lead to changed behavior outside the consulting room. But this is a common way in which a change of behavior takes place. And in most types of psychological therapy, the therapist uses this type of strategy in a deliberate way. An important part of therapy is exactly this: formulating rules and trying to increase the probability that the client will follow them.

BEHAVIORAL PRINCIPLES AND OTHER APPROACHES TO PSYCHOLOGICAL THERAPY

I have described two paths to change in the preceding discussion. Using typical behavioral terminology, you might say that in psychological therapies for verbally competent individuals, we establish consequences that reinforce new behavior in session and also train rule-governed behavior. The therapist can only act in relation to the client when they meet, in the first arena. There, the therapist can establish antecedents and consequences for behavior as it occurs. If new behavior occurs in the therapeutic situation and the therapist manages to establish reinforcing consequences, these can have lasting results on the client's behavior, provided that this new behavior meets with sufficiently reinforcing consequences out there in the client's everyday life, the second arena. In addition, verbal antecedents (rules) play an especially significant role, since they can influence behavior long after they have been established. And, for better and for worse, these rules entail a degree of insensitivity to direct contingencies.

In any psychological therapy rooted in behavioral principles, the therapeutic strategy is developed based on an understanding of the contingencies that influence behavior: antecedents and consequences. The therapist attempts to influence these contextual factors in different ways. Therapy is thus founded on a functional analysis, or ABC (Ramnerö & Törneke, 2008).

Since learning theory is about human behavior as a whole, it is, of course, possible to analyze all psychological therapies from this frame of reference, regardless of which theory the therapy in question is based on. If a therapist's actions in relation to a client in fact change the client's behavior, this must be occurring according to behavioral principles, if we assume that these principles are universally valid. When a psychoanalyst or a cognitive therapist achieves changes in a clinical setting (Leichsenring, 2005; Dobson, 1989; Beck, 2005) this takes place because their actions in therapy influence antecedents and consequences of the behavior that has been changed. There is no other possibility when we look at this from a behavioral perspective, regardless of whether the therapists themselves describe their work for change from a different theoretical angle. In the remainder of the book, I will describe a psychological therapy that is based on behavioral principles, with a special emphasis on conclusions drawn from relational frame theory. But first, let me outline a view we can take on two other important psychotherapy models as seen from this perspective. I do this from the conviction that behavioral principles are universally valid.

As I do this I am aware of two potential risks: First, there is a risk that this could be interpreted as presumptuous—that the claims seem too grand. That is an inevitable risk. The claims of learning theory *are* grand. This theory does not profess to explain only certain aspects of human behavior or a certain type of therapy. It takes as its starting point the view that behavioral principles are universally applicable, that these principles describe how we humans learn everything we learn and how it is possible to influence this learning process.

Second, there is a risk that some who advocate the therapy models analyzed here will feel that my descriptions aren't fair, since I will summarize these models and, to a large degree, will disregard the explanations offered by these therapies. All I can do in regard to this concern is clarify my own starting points. I am not attempting to describe a theoretical integration of the psychology of learning and other theories, nor do I intend to account for any form of dialogue among the theories of the different models. My intention is to give the reader a behavioral perspective on other models of psychological therapy. Such a perspective is possible; we can look at the central components of psychodynamic[1] and cognitive therapy from the principles of learning theory.

---

1 I am using the term "psychodynamic therapy" to denote all forms of psychological therapy based on what is usually called "psychodynamic theory" or "psychoanalytic theory." The borders between the different terms in use are not always distinct. Here, the term refers to the whole range of such therapies, from extensive psychoanalysis five sessions a week for many years to short manual-based therapies, for example, the kind described for so-called affect phobia (McCullough et al., 2003).

Behavioral Principles and Psychodynamic Therapies

It is not a novel undertaking to describe psychodynamic therapies from the perspective of learning theory. In several of his books, Skinner wrote about how the principles he deduced from his experiments can be used to understand how psychotherapy works (Skinner, 1953, 1957, 1974). The type of psychotherapy he analyzed is primarily psychoanalysis, along with humanistic psychotherapies. Other earlier writers have also analyzed psychoanalytically based therapies using learning theory (Dollard & Miller, 1950; Ferster, 1972).

Skinner noted two fundamental possibilities for exerting influence within the framework of psychoanalysis, and within psychotherapy in general (Skinner, 1974). The first is that the therapist, over time, develops a relationship to her client that makes her a source of reinforcement. The therapist receives the client's behavior in a different way than the client is used to in ordinary life. In the course of therapy, new consequences will become contingent on behavior. An important example mentioned by Skinner is that the therapist refrains from aversive (punishing) control. The experience of systematic aversive control influences people to use an avoidant behavioral repertoire. With such a history, people mainly act to diminish aversive events and get away. If the therapist behaves differently in this regard, the probability increases that avoidance is extinguished in the interaction between client and therapist and that the client tries out new behaviors, such as more clearly expressing his opinion about something the therapist has said or done. If the therapist then acts to reinforce this behavior, the possibility of learning new ways of relating to others has been established. The fact that sessions are frequent and the course of therapy continues for so long provides the therapist with many opportunities to establish consequences for the client's behavior that open up prospects for learning new behaviors. In frequent interactions that take place over an extended period of time, the therapist can control contextual factors that influence the client more than would otherwise be possible. Referring to the concept of the two arenas, you might say that psychoanalysis establishes the first arena to a very high degree compared to a therapeutic setting where therapist and client meet less frequently or for fewer sessions. But, of course, this is not automatically helpful to the client. Within the context of closely spaced interactions, the therapist must establish consequences to the client's behavior that actually enable changes the client desires. There is always the risk that the therapist isn't capable of this, and that, in the worst-case scenario, she actually strengthens or aggravates the type of behavior that is problematic for the client.

If we look at this situation from the standpoint of learning theory, a limitation of this method of exerting influence is that it requires generalization. If a client has learned a new behavior in interaction with his therapist and this behavior is not reinforced in the rest of his life, generalization probably won't occur. However, this limitation applies to all psychological therapy, since no therapist can control what happens outside of the therapeutic setting. That said, the frequent and long-term interactions that are a part of psychoanalysis provide many opportunities for influence.

Another possible limitation is that the behavior we want to influence needs to occur during interactions in therapy in order for it to be influenced by contextual factors that are present in the moment. Interpersonal behavior is, of course, highly likely to be cued in a framework like this. But problematic behavior in realms other than interpersonal behavior may only be prompted by external circumstances. Certain types of compulsive behavior and avoidance of specific external phenomena, like spiders or social settings, are examples of problematic behaviors that do not necessarily manifest within the setting of psychoanalysis. The actual behavior is not displayed, even though there may be an excellent possibility that the client will talk about it. However, the way the client talks about his problematic behavior may not be what needs to change. Since psychoanalysis usually emphasizes the importance of keeping therapeutic contact inside defined boundaries of time and location, this can obstruct establishing external situations that bring certain problems to the fore.

Skinner noted that one behavior that is often reinforced in a general way in psychotherapy is observation of one's own behavior—what is, within the psychology of learning, called self-discrimination. One part of self-discrimination could be learning to notice the connection between what you do and what happens when you do what you do. A client might say, "I noticed that she looked my way and I wondered if I should walk up to her to talk a bit, but I didn't do it." In a psychodynamic therapy, the therapist would be likely to reply, "Tell me more about this," or otherwise say something that reinforces continued self-discrimination by the client: discriminations of how it felt, what memories were recalled during the incident, what happened after that, and similar types of connections. This is close to what behavior analysis refers to as a functional analysis. If the client talks or thinks about these connections, this is verbal discrimination, which is an instance of the person tacting the contingencies of this specific situation. Although this is done in accordance with slightly different guidelines within psychodynamic therapies than within behavior analysis, it is still true that the behavior of tacting these contingencies is often reinforced.

This behavior—being able to tact one's own behavior and its connection to other events—is called insight, both in psychodynamic therapy and in everyday language. From the perspective of learning theory, it could be called outlook—which might be a more fitting expression (Ferster, 1972). You learn to discriminate, and, since in this case it is verbal discrimination, you learn to tact your own behavior as well as the connection between what you do and other events. Doing this is a prerequisite for what Skinner spoke of as the other possible way of exerting influence in psychotherapy in general, and in psychodynamic therapy specifically: formulating rules for behavior outside the therapeutic setting. Technically put, the therapist is to establish antecedents that increase the probability of a certain behavior taking place in a situation other than the one at hand. In other words, the therapist does something to increase the chances that a certain rule-governed behavior will occur in a setting other than in session with the therapist.

If this self-discrimination occurs as a result of therapy, the client will formulate rules like these whether or not the therapist actively helps him do so. In psychoanalysis the therapist usually gives this kind of help sparingly, but many other types of psychodynamic therapy make more room for mutual work to form such rules. Indirectly, this occurs in classical psychoanalysis too, since the client is a verbally competent human being and possesses the skills needed to create self-rules.

Clearly, these two essential ways of influencing the client's behavior can and do interact with each other. If a certain new behavior results in new and different consequences in interactions with the therapist, and especially if this happens repeatedly, this experience can form the foundation of a rule of this type: "If I also do this when I'm disappointed in Lee, he might listen." The person acts, meets certain consequences, tacts a contingency, and formulates a rule that specifies a possible behavior and a possible future consequence. If the person subsequently acts according to this rule, this is the type of behavior described as tracking in chapter 6.

We can describe different types of psychodynamic therapy, and other psychological therapies as well, based on the balance between these two strategies for working to change client behavior. In frequent and long-term psychoanalysis, the direct contingencies that are established in the interaction between therapist and client are very likely to play a large part in any changes. Therapy provides what some early interpreters of psychoanalysis called "a corrective emotional experience" (Alexander, French, & the Institute for Psychoanalysis, 1946, p. 66). Generally speaking, however, psychoanalysis has a strong emphasis on "understanding" and talking about the connection between actions, thoughts, emotions, and external events. The classic psychoanalytic interpretation can be understood as a verbally formulated discrimination of any such contingency. It is an example of how the therapist tacts such a contingency. In an ideal situation, an interpretation of this kind is a pure tact (described in chapter 2), though this is not necessarily so. This verbal behavior can lead to formulating rules, suggestions, or advice, or maybe simply to a description of different alternatives. Shorter and more prescriptive types of psychodynamic therapy have a stronger focus on influencing rule-governed behavior through helping the client formulate rules for alternative behavior. In principle, there is nothing to prevent therapies with less contact between therapist and client from having their main effects through direct contingencies established in the interaction between the two. However, the likelihood that this kind of contact contains new consequences that are effective and go on to become generalized in the client's daily life is not as high as when the therapeutic contact is frequent and of a long-term character. In therapist-client relationships that are short-term or involve less frequent contact, the therapist simply has fewer chances to establish direct contingencies for behavior. And yet frequent interaction is no guarantee that change will take place. Frequent and long-term therapeutic contact can consist of different types of verbal behavior without any contingencies for change being established.

From the perspective of learning theory, it is fair to assume that an ideal psychodynamic therapy has a balanced distribution of these two principles: nonverbal learning through direct contingencies that promote change, and formulating rules that lead to enhanced skills in tracking.

Behavioral Principles and Cognitive Therapies

Just as the term "psychodynamic therapy" encompasses slightly varying approaches, so does the term "cognitive therapy." The following discussion is based on Beck's definition: "Cognitive therapy is best viewed as the application of the cognitive model of a particular disorder with the use of a variety of techniques *designed to modify the dysfunctional beliefs and faulty information processing* characteristic of each disorder" (Beck, 1993, p. 194, italics added).

Cognitive therapy can also be analyzed using the two principles for exerting influence mentioned by Skinner (Zettle & Hayes, 1982). Using this book's terminology, you could say that the main goal in cognitive therapy is for the therapist to focus on influencing how rules are formulated.

The therapist initially spends time monitoring the relationship between the client's verbal behavior (thoughts, beliefs, life principles) and the issues the client describes as problematic. The therapist then uses a variety of techniques to question this verbal behavior and help the client reformulate it. An important part of this reformulation is achieved through different types of experiments. Based on what therapist and client formulate during sessions, the client is encouraged, in homework assignments, to check different hypotheses concerning the connections between his behavior and other events. For example, in dialogue with the therapist, the client may formulate a rule like "If I tell Lee my thoughts, he will despise me." Initially the therapist helps the client more clearly specify what this means. She may ask something like "How do you know he will despise you?" Other possible questions would be "Assuming that he were to despise you, how would that be made evident? Would you be able to observe it somehow? What would you interpret as a sign that he actually despises you?" Following this, she would help the client phrase an alternative hypothesis about how Lee might react and what indicators the client could try to observe that might point to this hypothesis being correct. Based on this dialogue, a homework assignment is constructed in which the client checks whether this hypothesis is in line with how Lee actually behaves (Beck, 1976). The therapist thus helps the client tact connections between his own behavior and other events; in this case how Lee reacts if the client tells him certain things. The homework assignment could become a way of training the client's tracking repertoire.

Clearly, cognitive therapy places a strong emphasis not just on how rules are formulated, but also—through these types of real-life experiments—on arranging for the client to contact direct contingencies by way of how these things actually turn out. The client then brings these experiences back to the conversation with his therapist, where they are used as a basis for new formulations, which in turn can be checked.

When behavior principles serve as the basis for understanding the outcome of therapy, a limitation of this therapy model, and all models of psychotherapy conducted primarily in outpatient settings, is that the therapist is not in control of the consequences the client encounters outside of the therapy setting. In the example above, the therapist doesn't control how Lee actually reacts. And in order for therapy to effect change in the long run, generalization is required. This requires that the client actually encounters reinforcing consequences for the new behavior he is experimenting with outside of the therapy sessions.

How, then, does cognitive therapy allow for direct reinforcement of behavior in the ongoing interaction between therapist and client? One kind of behavior is almost assuredly reinforced: the verbal behavior of formulating tacts and, in time, rules for behavior. You learn—partly through direct contingencies—to talk and think in a certain way. But as mentioned in regard to psychodynamic therapy, this is not necessarily of help to the client. The way the client thinks and talks about his behavior may not be the behavior that needs to be changed. But to the extent that new rules for behavior are formulated, these can function as antecedents for behavior in different situations and thereby lead to new types of rule-governed behavior that work better for the client.

Another feature of cognitive therapy that increases the therapist's chances of influencing direct contingencies in connection with the client's behavior is engaging in behavioral experiments while the therapist is present. These experiments are set up in much the same way as the homework assignments are. Let's say the client formulates the rule "If I feel too much anxiety, I'll lose control completely." In this case, the therapist first helps the client specify precisely what the different parts of the rule mean: How much anxiety is too much? What would be an example of losing control completely? Then therapist and client create an experiment aimed at checking the core content of the rule. In this case, the experiment should aim at increasing the client's level of anxiety. How this is done depends on what external situation is most likely to make this person anxious. It could be handling knives, going through a car wash, or doing something unusual in full view of other people. Even in these situations, where the therapist participates in the experiments, she is not in control of most of the consequences that occur as a result of the client's behavior. But there are usually fewer variables that are out of the therapist's control than in homework assignments she doesn't take part in. In addition, the therapist also has the opportunity to observe behavior that occurs in the moment—for example, behaviors triggered by the client's experience of anxiety—and then reinforce alternative behavior in the moment. The client's experiences from the experiment are then used to formulate prospective new ways of behaving. Therapist and client tact the contingencies of a certain behavior and its results in the experiment and formulate tracks that in turn hopefully increase the probability of new behavior in future situations. From a behavioral perspective, these types of experiments should be an ideal way of practicing the repertoire of tracking. And indeed, leading cognitive therapists regard these types of experiments as essential to cognitive therapy, especially in the treatment of anxiety disorders (Clark et al., 2006).

The work with direct contingencies in the therapist-client interaction that is a central part of psychodynamic therapy also occurs within the framework of cognitive therapy (Safran & Segal, 1990), and behavioral principles do a good job of explaining this kind of therapeutic work. When the client's problematic behavior is triggered in the first arena, in his interaction with the therapist, her responses can then function as reinforcing for alternative behavior. Additionally, by making the connections between the client's behavior and other events the subject of a joint analysis, she facilitates the formulation of rules. These rules can subsequently function as antecedents for new behavior outside the therapeutic setting, in the second arena. Currently, this way of using the therapeutic relationship as a focus in therapy occurs across the boundary between psychodynamic and cognitive models of psychotherapy (Safran & Muran, 2000; Newman, Castonguay, Borkovec, Fisher, & Nordberg, 2008).

Different models of cognitive therapy can thus also be analyzed based on the balance between the two possible strategies for change described by Skinner (direct contingencies and establishing new rule-governed behavior). This is especially clear if we include the models that are closer to the behavior therapeutic tradition in the discussion. Beck's model of cognitive therapy was influenced by behavior therapy at an early stage, especially when it comes to therapeutic techniques; therefore the strategies that are meant to provide the client with direct corrective experiences have always held an important position within cognitive therapy (Beck, 1976, 2005). Within cognitive therapies there are also methods that assign less importance to this and instead emphasize the importance of influencing the client's ways of thinking and speaking about his behavior and its relationship to other events. The balance between the two strategies probably also varies from therapist to therapist. There is, however, a current trend toward stressing the importance of therapies offering components that, in behavioral terms, influence the client through direct contingencies. Parts of this trend are closely linked to an increase in the influence of behavior therapy (Farmer & Chapman, 2008), but the trend can also be seen within variants of cognitive therapy where this influence is not as pronounced. In schema therapy, for example, a type of interaction between therapist and client called limited reparenting is recommended (Young, Klosko, & Weishaar, 2003). In this approach, the therapist acts in a way that lets the client encounter consequences of his behavior that differ from his experiences early in life. This approach also calls upon the therapist to respond to some of the client's needs that haven't been met in his earlier experiences, within the appropriate boundaries of the therapeutic relationship. The goal of this strategy is described by advocates of schema therapy using Alexander's term from 1946: The client is to have a "corrective emotional experience" (Alexander et al., 1946, p. 66).

PSYCHOLOGICAL THERAPIES AND A NEW UNDERSTANDING BASED ON RFT

From a behavioral perspective, learning new things in the context of psychological therapy takes place in two fundamental ways: First, new direct contingencies are established in therapeutic interactions. And second, new rules are established that increase the probability of changes in rule-governed behavior. The analysis in this chapter makes it clear that these two paths to change are used in therapies that are not explicitly based in behavioral principles. What we hope to do through therapy that does take its starting point in a behavioral understanding is provide something that, in the long run, demonstrates a greater capacity to help people precisely because it is deliberately founded on this specific scientific analysis of change processes. Therapy models that are explicitly based in behavioral principles have displayed promising results, even when they haven't integrated the new knowledge from the field of relational learning. Two examples of this are dialectical behavior therapy and behavioral activation.

When these two paths to change (establishing new contingencies of reinforcement in session and establishing new verbal antecedents for behavior outside of session) are described as different from each other, it is important to understand that they are different only in certain respects. Naturally, they are also similar. They are alike in that both involve a therapist who works to alter contextual factors in the present moment and thereby influence behavior. Relational framing is influenced by contextual factors in the present moment, just as all other behavior is. There is no other way of exerting influence. Arbitrarily applicable relational responding does not take place in a different realm than other behavior does. It is something an individual performs in the here and now, in a specific context. What separates the two paths to change is that once a person has learned relational framing, his behavior can be influenced in a way that wouldn't be possible if he had not learned this ability. This is due to the way in which stimuli can now acquire their functions for an individual.

What contribution is made by this understanding of relational learning? First, it means that the scientific analysis of behavior that has been evolving over the past century can now more effectively take on phenomena that this tradition once had difficulties dealing with. The functions of human language and cognition are now more readily available for behavior analysis, and insofar as these are important factors in connection with psychological problems, we can benefit from this increased understanding. Phenomena that have already been subject to psychological treatment from other theoretical approaches are now also available to a scientific method that has proven very successful in fields other than what has traditionally been called psychotherapy (Ayllon, Layman, & Kandel, 1975; Peterson & Azrin, 1992; Sallows & Graupner, 2005). The understanding of relational learning offers the hope that clinical behavior analysis can take a general step forward in working with verbal behavior. The latter has already begun within certain types of cognitive therapy (Ellis, 2005; Blackledge et al., 2008).

The second contribution to clinical behavior analysis as a direct result of our increased understanding of relational learning is what was discussed in chapter 7: the side effects of human language. Beyond supplying us with increased flexibility, rule-governed behavior also opens the door to a number of traps. These traps are central to common clinical problems and thus are of immediate relevance to clinical therapies. Changing rule-governed behavior is at the core of all psychological therapies. But if rule-governed behavior in itself is part of the problem, one of the therapist's most commonly used tools could function as a Trojan horse. All psychotherapy is talk therapy, more or less. If language is part of the problem, how do we go about using it to bring about change? Could it be that, in fact, psychological therapies need to offer some kind of release from the power of language, at least in some respects? This seems to be a reasonable conclusion from the research upon which RFT is built. And this conclusion is, for the most part, something new.

As the experimental work on relational conditioning has proceeded over the last twenty years, clinical interventions have been developed based on these findings. One result is a particular form of clinical behavior analysis: acceptance and commitment therapy (ACT, pronounced as one word). This model integrates behavioral interventions that have long been established—interventions like exposure and behavioral activation. It also places a strong emphasis on interventions that are motivated by the new knowledge summarized in part 2 of this book. This especially applies to the knowledge that, in psychological therapies, it is often essential to diminish the dominance of problematic verbal behavior over other human behavior.

CLINICAL BEHAVIOR ANALYSIS

The remaining chapters of this book are devoted to clinical behavior analysis, with a particular emphasis on how it can be reshaped in light of our knowledge about relational learning, including its dark side. Since ACT has been developed with precisely this agenda, this model of therapy will predominate. However, my aim is not primarily to introduce ACT, but to describe clinical behavior analysis in a broader sense and how our knowledge about relational learning develops and deepens this entire tradition. I will therefore describe clinical interventions within a structure that more closely resembles classical behavior analysis than the specific model often used in presentations of ACT (see, e.g., S. C. Hayes & Strosahl, 2004; Luoma, Hayes, & Walser, 2007; Bach & Moran, 2008).

I do this for several reasons. First, I want to clearly illustrate the immediate connection between a classical functional analysis and the interventions I propose. The purpose of this book is to show how learning theory, including more recent findings in this field, is directly applicable in clinical practice. An argument that has been given for developing the special description of ACT that dominates most published texts about the model is that it is meant to facilitate learning by therapists who are not very familiar with learning theory, so they can readily make use of the ACT model in their work (S. C. Hayes, 2008). Since I, quite to the contrary, intend to describe the behavioral foundations, this argument does not apply here. Second, I want to stress that the new interventions recommended in ACT are not defined as belonging to any specific model of therapy. Learning theory is a scientific theory and its practical applications are freely available, rather than being associated with any specific model of therapy. Last but not least, I want to describe a psychological treatment that is based on principles rather than well-defined manuals. The therapy models that have their foundation in behavior analysis all underline that they are governed by principles rather than by manuals. This goes for acceptance and commitment therapy and functional analytical psychotherapy, as well as for behavioral activation. And yet these approaches are often described precisely as if they were distinct, delimited models, rather than simply different aspects of the same therapeutic tradition—and this despite the fact that they all openly admit their close affinity. Consequently, different attempts have been made to integrate the different models (Callaghan, Gregg, Marx, Kohlenberg, & Gifford, 2004; Kanter, Baruch, & Gaynor, 2006; Kanter, Manos, Busch, & Rusch, 2008). These attempts usually involve adding aspects of one model to one of the other models. In the remainder of the book, I want to take a more fundamental approach. I intend to describe clinical behavior analysis as one integrated therapy model, based on the principles of behavioral psychology and with a particular focus on the contribution offered by RFT. This is not done from a position of arguing with any of the individual models or with the more common ways of describing them. My intention is to highlight their common ground and point to the possibility of a coordinate application.

The structure I will follow is taken from functional analysis. The therapist can, in a given moment, focus her interventions on influencing either the antecedents or the consequences that surround a specific behavior. In the next chapter, I will describe some common starting points of clinical behavior analysis. In chapter 10, I will go on to describe interventions that focus on consequences, and in chapter 11 I'll look at interventions that focus on antecedents. In all of these chapters, a key focus will be on interventions that have their basis in RFT.

CHAPTER 9

General Guidelines for Clinical Behavior Analysis

The therapeutic setting of clinical behavior analysis is the same as it is for psychotherapy in general. The therapist meets with the client for an hour or two, with a certain number of days or weeks between sessions. These conditions are further tailored to each individual case. If the therapeutic work concerns difficulties that are generalized across broad areas of the client's life, such as in chronic depression or what is usually categorized as personality disorders, therapy may continue for several months or sometimes several years. Therapies for treating specific phobias, on the other hand, can often be conducted over just a few individual therapy sessions that are fairly close together in time, and perhaps of a longer duration per session. But it is not just the type of problem that determines the time frame of treatment. The goal of therapy is naturally an important factor. A person with a very high degree of functional impairment or a diagnosis that is generally seen as serious may be in therapy for a short period of time because the goal is limited. For example, a person diagnosed with chronic schizophrenia may be attending therapy with the limited goal of being able to go about most of his daily business despite his troublesome auditory hallucinations.

Clinical behavior analysis is also subject to the general conditions of psychological therapy described in chapter 8. All influence on the client is exerted in the moment in two primary ways: by establishing direct contingencies, and through rules. When it comes to direct contingencies, setting up new consequences of behavior is key. Rules are simply verbal antecedents, and they are important because they can govern behavior at much later points in time and in different contexts than the situation in which they are given. Another important aspect of rules is that the behavior they govern is relatively insensitive to direct contingencies.

INITIAL CONTACT AND THE TWO ARENAS OF THERAPY

The therapist only has direct access to the arena she shares with the client, that is, whatever takes place when the two of them meet. This means that any influence must be exerted in this arena. At the same time, the client is seeking help in regard to things that happen in the second arena: his life outside of sessions with the therapist. The first arena is used to influence the second arena.

The client usually focuses on the second arena during initial contact with the therapist: He relates his problem and how it manifests in his life. Yet his problem may also present itself in the first arena even during this initial contact with the therapist. If the client describes his problem as being that he worries too much, he may also be able to report that he is worrying in that moment—about, for example, how therapy is going to turn out. Sometimes the client may not report this, but the therapist feels she can observe it. Perhaps the client asks several questions about the treatment and does so in a way that seems to be precisely "the client being worried." Based on the fundamental condition of therapy—that any and all influence takes place "now"—it is important for the therapist to note any signs of the client's problematic behavior occurring in the moment. This applies to the initial contact and throughout the entire therapeutic interaction. In fact, these occasions provide especially favorable opportunities to exert influence. Still, most therapies begin with a focus on the second arena. Most people's expectations of psychological therapy lead them to believe this is natural. It is also natural from the therapist's perspective, since the client's story often provides an overall picture of the problem.

However, it is important for the therapist to remain aware that when the client relates something, what is directly accessible is the client's way of talking, not the actual phenomenon he describes. As it is, what is demonstrated is verbal behavior. If verbal behavior is part of the problem, then there is an embedded problem—a problem inherent in the conditions of therapy, as all psychotherapy is "talk therapy." This gives the therapist further reason to make use of the occasions when the more direct problematic behavior is demonstrated in the first arena. Another common way in which behavior therapists try to circumvent this dilemma is by using different rating scales, logs, and questionnaires in the hope that this will provide more direct contact with what occurs in the second arena. While this is an important contribution, filling out rating scales is, of course, not the same behavior as that the scales aim at describing. There is no getting around the quandary that, at the beginning of a therapeutic relationship, what the therapist faces first and foremost is verbal behavior. This illustrates a point I made in chapter 3: that an increased understanding of verbal behavior is incredibly important in psychological therapies. Talk therapy needs a language theory! So let's take a look at what we need to do in the initial phases of a therapeutic interaction based on the theoretical view presented in this book.

PERFORMING A FUNCTIONAL ANALYSIS

The fundamental element of behavior analysis is describing behavior based on ABC (Ramnerö & Törneke, 2008). A certain behavior (B) is to be understood and influenced. This is done by identifying the contextual factors (A and C) that influence the behavior in question. Following this, the therapist attempts to alter these contextual factors. In the initial conversation with a client, the description in itself is focused. What behavior is to be influenced, and which antecedents and consequences are available and useful in this process?

In experimental behavior analysis, these factors are immediately observable and available for manipulation. The analyst is present in the relevant course of events and has access to the relevant contextual factors. To some extent, the same goes for applied behavior analysis if the therapist is present in the life situation where the behavior takes place. This is the case in institutional therapy, such as in schools and inpatient treatment. But the circumstances are different in clinical behavior analysis. What I have called the second arena is accessible only through what the client relates; through his verbal behavior. This means therapy in this setting takes its starting point in verbal behavior. The fact that this can still change the client's behavior in the second arena as much as it evidently does is an example of how powerful verbal behavior is. Psychotherapy clearly has an impact (Roth & Fonagy, 1996).

So what are we trying to do through functional analysis under these conditions? We are trying to discriminate the type of behavior that is problematic for the client. In behavior analytic language, you could say that we are tacting behavior and different relevant relationships between behavior and contextual factors. We do this for two purposes. First, the therapist is attempting to identify contingencies in the client's story that might be used in interventions for change in the long run. What is the client doing that results in unwanted consequences? The therapist assumes that what the client relates corresponds sufficiently to what actually happens, and that this narrative is thus of use in therapy. Such is the therapist's approach even though she is undoubtedly aware that this may not be entirely the case. Clinical behavior analysis starts from the assumption that the client is tacting his own behavior in a sufficiently pure way, at least with the help of the therapist and her supportive questions. The client is naturally a firsthand witness to his own life and experience. However, given the shortcomings in this starting point, it is important for the therapist to constantly have the second purpose in mind: identifying any behavior she thinks may be problematic if it shows up in the present moment of the interaction. This makes the therapist a witness, too, at least to some aspects of what occurs for the client. The best conditions for a useful functional analysis are in place if it is based partly on the client's story from the second arena, and partly on something that is taking place in the first arena, where the therapist is present.

The following client-therapist interaction demonstrates how this can be done. (This is a typical but fictionalized interaction, as are all case examples in this book.)

Anne says that she has always been a worrier. For many years it wasn't very troubling; she just thought it was part of her nature, but this changed in a substantial way after she traveled through Southeast Asia four years earlier. While walking along a beach late at night, she was robbed by three men, who appeared behind her suddenly, knocked her over, and made her give them her camera and all the money she was carrying. They had a generally menacing attitude, but they didn't harm her. After this episode she went back to the hotel and met the friend she was traveling with. When she started to tell her friend what had happened, Anne experienced intense anxiety and dizziness and, as she reported, "turned totally weird." When these symptoms did not subside after a couple of hours, her friend took her to a hospital, where she stayed for a day and a night. They told her nothing was wrong with her and gave her an injection that helped her relax.

After arriving back home, Anne was troubled by anxiety attacks in connection with things that reminded her of the event. Over a period of two months she met with a psychologist, and her troubles abated. She doesn't think about what happened on her trip a great deal any more, and it doesn't feel painful for her to think about it. However, Anne continues to feel that she has changed. She feels curtailed in her day-to-day life, and it seems to her that she has been injured by what happened. She says, "I am not my usual self any more," and reports that a number of troubles in her daily life trigger these types of thoughts: physical tension, tiredness, and particularly the experience of not being able to concentrate. She devotes a lot of thought to what might be wrong and what she should do. Periodically, she feels low. She thinks about how her life has been ruined and believes that what happened during her holiday trip caused this injury. She has intervals when she feels good, but then the same feelings return. She has seen medical doctors and gone through different examinations in order to eliminate physical problems, but each time she is told that nothing is wrong with her.

After Anne has given the above account and different examples of how she gets mired in brooding in everyday situations, the following dialogue with the therapist takes place.

Therapist: This trouble you mentioned—the "weirdness"—can you feel any of that now?

Anne: Now, when I'm here with you, you mean? Yes, I actually do feel some of it now. It's like I can never get rid of it. It's hopeless! Something went wrong when that thing happened—something in my brain. And still, I don't think about what actually happened that much. It happened and it was terrible, but it feels like a long time ago. I got to talk about it when I came home. But something went wrong inside me. And I guess the doctors don't seem to be able to find it or do anything about it. What could it be? It just fouls up my work completely. I can't think straight. Where will all of this lead? I was offered a new position last month, but I don't know if I have the guts to accept it. What if I can't handle it? What if this never stops?

Therapist: Is this what happens to you?

Anne: What do you mean?

Therapist: I mean, what happened right now?

Anne: That I feel some of it, you mean?

Therapist: Yes, and even more specifically, what you seem to do when you feel it. It's like there are two parts in what happens, and it seems to be happening right here and now. First you feel some of the "weirdness," and then you start reasoning and asking questions: What's wrong? Where will this lead? and so on. It seems like you're trying to sort it out. Or am I misinterpreting you?

Anne: No... Okay... Yes, I see what you mean. I don't think I've seen it that way before, but now that you say it, I think you're right. First something reminds me of it—again—and then I just kind of dive right in there. Yeah...that's exactly what usually happens.

Note how the therapist uses what Anne has told her about her problems. When, after hearing Anne, she sees something in their interaction in the room (the first arena) that is similar to what Anne is relating from the second arena, the therapist focuses on this in her analysis. Together with Anne, she tacts the different parts in an ABC analysis of the behavior taking place here and now, then compares this with what Anne can observe in her daily life.

WHAT BEHAVIOR IS TO BE FOCUSED?

At the beginning of the client-therapist relationship, the therapist scans the client's story as he relates it. Naturally, her knowledge of how problems usually manifest themselves plays a part. The therapist has assumptions of her own. All the while, it is important to take a broad view of what the client reports. The starting point is the client's problems and how they show up in his current situation. Events from earlier times can also be used, especially as additional examples of things that are currently in focus. Events from long ago can, of course, be very important to something that is taking place much later, and can even be central to the client's problems. However, it is difficult to use events from the distant past as the sole basis for a functional analysis, as all of us have poor direct access to these kinds of events and processes. This makes it difficult to achieve the kind of detailed description we hope to arrive at with a functional analysis. Furthermore, the problems that have led the client to seek help are closer at hand. People who had problems many years ago but don't currently experience any difficulties don't usually seek psychological help. What is needed for a functional analysis is always in closer proximity. However, what we commonly call "memories" are still interesting in terms of the function they have now for current behavior.

Initially, the client's problematic behavior is at the center of attention in the therapeutic interaction. What is this person doing that contributes to things not turning out the way he wants them to? In the literature, this is often called "a behavioral excess" (Kanfer & Saslow, 1969), something a person does too much of. Some problematic behavior is not supposed to be performed at all. Abusing children is a simple example of this type of behavior. But most of the time, problematic behavior is simply about doing too much of something. Anne's example, in the dialogue above, is typical in this regard. She asks herself questions, she thinks about them, and she tries to sort it all out. No one would suggest that this is something that she should stop doing altogether. This behavior is very useful in many situations and is even vital to us humans. But Anne does too much of this behavior, and she uses it in contexts where it doesn't work.

The other type of behavior that gradually becomes important to tact in the therapeutic setting is alternative behavior. This is behavior that, when performed, increases the probability that things will turn out as the client wants them to—behavior that increases the workability of life. Alternative behavior can often be described as a deficit: behavior that exists in the client's repertoire but is seldom used, or at least not used in the context where it would increase the workability of life. Of course, it may be that a desired alternative behavior doesn't exist in the client's repertoire. There could be someone who has never—in any situation whatsoever—put his foot down to someone in a position of authority, for example; someone who reports that he "doesn't even know how to go about something like that." In many cases, however, this description is actually a way of saying, "I don't want what I think that would entail," rather than a true indicator of a complete lack of capability.

In principle, then, there are two types of behavior in focus in a clinical functional analysis: problematic behavior and alternative behavior. We could say that we have two types of clinically relevant behavior. This brings us to a point where RFT can cast light on what the therapist should be especially on the lookout for.

EXPERIENTIAL AVOIDANCE AS THE FOCUS OF FUNCTIONAL ANALYSIS

A core function of most problematic behaviors that dominate when people seek psychological help is trying to avoid things. Problematic behavior is often governed by negative reinforcement. The person does what he does because that helps him get rid of things he does not want, at least in the short term. These behaviors are under aversive control. As described in chapter 7, the capacity to frame relationally opens the door to a very wide interface with pain. Private events like thoughts, feelings, and bodily sensations can acquire aversive functions in themselves and thereby become the object of avoidance efforts. Rule-governed behavior, and the problem-solving strategies it results in, can easily develop into behavioral excesses, as in Anne's example. When Anne experiences thoughts that something is wrong or that something isn't going to work out, indirect stimulus functions dominate—functions that she accesses through arbitrarily applicable relational responding. Anne doesn't act based on simply having a thought; she acts based on the content of the thought, or what we might commonly refer to as its meaning. She acts on the basis of a story that, among other things, deals with the future. You might say that instead of regarding this story as something that is taking place within her in the moment, she regards her future life from the perspective of this story. She "dives right in," to use her own term. If we were to inquire more closely about the point of this behavior, Anne would probably describe it as an effort to dissolve the threatening aspects of the story, either by "understanding what it's about" and thereby being able to thwart the threat, or by realizing that "it's nothing dangerous." There is nothing inherently unnatural or strange in this behavior. This is what we all do when we face threats. Anne's problem is that her experience of this threat is distributed throughout her life, and the resulting behavioral strategy dominates and doesn't actually work well in regard to what she wants to achieve. When those bodily sensations return, her thoughts about what this means return, as well. She cannot get rid of them. Her attempts to get rid of these experiences actually constitute the type of behavior that adds to her problems and restricts her life.

Anne's behavior is an example of experiential avoidance. She is occupied with these attempts to eliminate or control phenomena within her experience that are not under voluntary control. Because her therapist knows that this is a common behavioral strategy, she has reason to look for exactly this. Behaviors that vary widely in external qualities can nevertheless serve the same function of experiential avoidance. Although the behaviors are topographically different, they are functionally alike since they serve the same purpose. They belong to the same functional class. You can try to get rid of unpleasant feelings, painful thoughts, or distressing recollections in different ways: In a face-to-face interaction, you might look aside and fall silent. But you could also talk energetically or raise your voice, if your purpose in this is the same: ridding yourself of the unwanted experiences. In a situation like Anne's, another person might try to think of something else. Anne, on the other hand, thinks about it more. These topographically distinct behaviors can be functionally alike if both aim at getting rid of the unpleasant experience. As another example, when people experience grief, both increased activity and passive withdrawal can be aimed at reducing the feelings of sorrow and loss.

FUNCTIONAL ANALYSIS IS THERAPY BEGUN

What therapist and client do in a functional analysis is mutually tact the client's behavior and the relations between this behavior and the contextual factors that influence it. This brings us back to Skinner's statement that "a person who has been 'made aware of himself' by the questions he has been asked is in a better position to predict and control his own behavior" (Skinner, 1974, p. 35).

How tacting one's own behavior, and the contingencies that influence it, can at a later point alter the kind of behavior that is being tacted is explained by RFT. Verbal behavior, such as words or thoughts, can alter stimulus functions of other events through mutual entailment and combinatorial mutual entailment. Take Anne and her effort to eliminate the feeling that something is wrong with her. Her behavioral starting point has been that if she thinks more, she will reach what she is striving for: somehow eliminating the threat. This does not necessarily mean that she's had this in her thoughts explicitly, but she has been acting from such an assumption. She has behaved as if such an approach is the most viable way forward. In the following interaction, the therapist uses a functional analysis in an attempt to affect the stimulus functions of the behavior of "diving right in."

Therapist: So, first, something reminds you. This time I guess it was my question. And then you dive right in there... What do you feel is the point in that? The point in diving in there?

Anne: Point? I don't know. What am I supposed to do? I have to try to solve it somehow, you know.

Therapist: So you could maybe say that this "trying to solve it" is the point. That's what it's all about: solving it.

Anne: Yeah, you could maybe say that. There has to be a solution, you know.

Therapist: Yes, that sounds like a reason for diving in there, I think—to find a solution. Is it working?

Anne: *(silent for a moment)* It's not working at all. The more I think about it, the worse it gets.

Therapist: What if your experience is telling you something important here? "Diving right in" introduces itself as an obvious road forward. And when you do this, it just gets worse. Maybe "diving in there" is what doesn't work?

Following this dialogue, the behavior "diving right in" may acquire new stimulus functions for Anne. Previously this behavior had seemed to be the only possible behavior—what she had to do to find a solution. After the dialogue above, it acquires the function of "something I do that makes things worse." Of course, this isn't necessarily enough to spark a change, but the very analysis of the relationship between behavior and consequences can influence future behavior, and from the viewpoint of RFT, we can understand why this is so.

In ACT, this part of a functional analysis is used to generate what is called creative hopelessness. The therapist helps the client contact the connection between what she does and what the actual consequences are. If the consequences are something the client doesn't want, functional analysis transforms this course of events into creative hopelessness: hopelessness in the sense that the strategy the client is applying doesn't work, and creative in the sense that precisely this experience of how hopeless the strategy is can open the door to new alternatives.

CAPTURING BEHAVIOR

In the example with Anne, the therapist captures something the client does in the moment, as a natural part of the interaction. The therapist brings the focus to something the client does in the moment, which then becomes the subject of dialogue and intervention. In Anne's case, this happens early in the course of therapy, but it often takes more time for such a process to become clear to the therapist. In any case, it is important for the therapist to constantly remain alert to this possibility. These situations where clinically relevant behavior is occurring are useful in two ways: The therapist can use the behavior as the basis for a functional analysis performed together with the client, and she can also intervene for change in other ways.

Functional analytical psychotherapy places a special emphasis on this type of analysis of interpersonal behavior as demonstrated in relation to the therapist and describes many different ways to use this kind of interaction to facilitate behavioral change (Kohlenberg & Tsai, 1991; Tsai et al., 2008). However, as the example with Anne demonstrates, not all relevant behavior is interpersonal, strictly speaking. It is true that Anne's problematic behavior is demonstrated in the dialogue with the therapist here, but in the second arena, Anne's daily life, she does more or less the same thing, but with herself as the listener.

Many times it is ambiguous to the therapist whether a problematic behavior is playing out in the interaction. When the client's story about what happens in the second arena describes what the therapist interprets as a problematic behavior, the therapist can seize the opportunity to simply ask the question. Richard's story provides an example of this.

Richard has felt dejected for years. He says he has perhaps always been "a bit on the gloomy side." He works as a teacher and thinks that his performance has gone downhill gradually over the past year. He has periodically been away on medical leave with "depressive burnout" and has been taking antidepressants, but they haven't helped appreciably. He says that he actually likes most of the things he does in his work, but that he isn't given the most appealing tasks any longer, reporting, "The young overachievers get all of them." He ends up in the background at home, too. He and his wife, Erin, have two children in their early teens. Erin is active and devoted, both in her own work and in their family life. Richard feels like he's falling behind. He tags along with Erin's initiatives but finds himself more and more passive. His family has started to complain that he keeps digging in his heels and that they can't count on him. In his dialogue with the therapist, Richard has a rather docile attitude. He answers questions and is mostly attentive, but he does seem dejected. The following dialogue takes place during his fifth session.

Richard: I feel burdened. Whatever I do, it doesn't make any difference. It's like a haze. I manage to get through the days, but nothing ever seems to change. See, there's just no strength.

Therapist: When you feel burdened like this and there's no strength, what do you do?

Richard: You mean at work?

Therapist: Yes, as an example.

Richard: I try to do what I can. You have to take one thing at a time. It's no use getting burnt-out again, you know. I can remember when things were worse, but it's not much fun this way either. I'm just hanging in there.

Therapist: And what do you do at home?

Richard:	I guess it's the same thing. I'm feeling more and more burdened at home too. But Erin keeps it up, and I guess I tag along.
Therapist:	When I hear you say you're tagging along, it sounds rather passive, at least to me—like "lying low." Does that sound right to you?
Richard:	Yes, sure. I don't know what to do. There's no strength left.
Therapist:	Okay, there's no strength and you're lying low, but still hanging in there.
Richard:	Yes, so far...
Therapist:	Is "lying low" something you think you're doing here?
Richard:	What do you mean, "here"?
Therapist:	Here with me, in our conversations. I've noticed that you listen to me, that you answer questions, and that you participate in what we're doing here. And at the same time, I wonder if somehow you're lying low here, too—as if it's no use here either, as if you're acting the same way here as you do at home or at work.
Richard:	*(after a moment of silence)* Yes, now that you mention it. I mean, I want this to help; I want it to get better. But I think I'm lying low here too. I do. I guess it's something that comes naturally to me, maybe.

EVOKING BEHAVIOR

A lot of problematic behavior can be captured in the interaction between client and therapist. In my experience, more so than may seem to be the case. But there is also a need for the therapist to evoke client behavior. We need to access the client's behavior in a direct way, beyond what is demonstrated or triggered by the questions in a dialogue. This is necessary to perform a functional analysis and to intervene for change more directly.

Behavior therapy has a long tradition of doing precisely this. The term usually used for this type of therapeutic work is "exposure." This is also the single psychological intervention most widely supported by scientific data (Barlow, 2002). It is used in all types of evidence-based therapies, most typically in treating anxiety disorders. While the strategy is simple, this does not necessarily mean that accomplishing this type of intervention is easy. On the contrary, it can be rather difficult and require a high degree of competence in the therapist. In exposure, therapist and client deliberately seek the type of conditions under which the client's problematic behavior usually occurs. In treating a specific phobia, like fear of injections, therapist and client arrange a situation where an injection is to be given. In treating compulsive washing, the therapist and client might go to visit bathrooms. In treating post-traumatic stress characterized by painful memories, the client is asked to recall what is tormenting him. All of this is done precisely for the purpose of evoking behavior. The aim is to allow the therapist to be present when the client's problematic behavior occurs, partly to help the client recognize what he is doing that doesn't work, and partly to influence the client in finding a response that works better for him. Exposure therapy doesn't focus solely on the client being exposed to aversive phenomena; it also looks at what the person does when he is exposed. Being exposed to the problematic situation is usually a necessary part of learning something new. It is in this situation that the person can learn alternative behavior, and also learn to recognize it so that it might be integrated into his behavioral repertoire. We might therefore say that skills training is a part of exposure therapy.

Deliberately evoking behavior in the arena of therapy has a wider application than you might think when you hear the word "exposure." It is everything that the therapist does, often in agreement with the client, in order for clinically relevant behavior to occur in this arena. For example, if bulimic behavior is the focus of therapy, deliberately evoking behavior might mean arranging for the client to have a meal just before the therapy session. For a client with a different problem, it might involve the therapist sitting silently and adopting a neutral waiting attitude if the client's problematic behavior is triggered under these types of conditions. Different types of role-playing can be used to evoke what the client needs to work with.

Within ACT, a number of experiential exercises are used to evoke or elicit behavior (S. C. Hayes, Strosahl, & Wilson, 1999). They are used both to discriminate problematic behavior and to help the client find alternative ways of behaving. In addition to using exercises that have been specifically developed from the theoretical principles ACT is founded on, the therapist is also encouraged to create new exercises that fit the moment and are based on the same principles. Some of these exercises are more like what is traditionally called exposure, in that they aim at establishing a situation where the client can be expected to experience negative affect. This is natural, considering that experiential avoidance is seen as a major problem. One such exercise involves the therapist and client placing their chairs in close proximity, then silently looking into each other's eyes for a couple of minutes. This exercise utilizes the fact that most of us feel uncomfortable when what we think of as our personal territory is infringed upon. And yet there is no real threat at hand. The instructions for this exercise are simply to sit still in this way and note how it feels—to note the different thoughts that come up and any other reactions. Afterward, therapist and client talk about what was noted, including thoughts, feelings, and bodily sensations. The client may have noticed an impulse to cut the exercise short. In that case the conversation can revolve around what it was like to just sit and observe this. Or maybe the client did cut the exercise short, providing an opportunity to talk about that. If the client's problem is experiential avoidance, the conversation will be about the two types of clinically relevant behavior—problem behavior and alternative behavior.

The following dialogue with Anne illustrates another exercise that is also aimed at discriminating these two types of behavior. Or perhaps, as in the eye-contact exercise, it is aimed at behaviors that are in some sense analogous or functionally similar to what the client has described.

In her dialogue with the therapist, Anne has described the difficulties she's trying to manage: physical symptoms, lack of concentration, and alarming thoughts. She describes her own efforts, how those efforts don't really work, and also her frustration at not being able to find an alternative.

Therapist: There's an exercise that I use sometimes. Would you be willing to try it?

Anne: Sure.

Therapist: *(brings out a notepad)* I want to make a note of some of the things you think are the worst problems, the things that bother you and that you said you want to get rid of—your lack of concentration, for example *(jots down something on the notepad)*. What else is essential?

Anne: The tiredness, the tension in my body...

Therapist: *(writes what Anne mentions)* How about the thoughts about not being your usual self and the risk that things are never going to get better?

Anne: That's the worst part—that it's going to go on like this.

The therapist notes the things Anne tells her and then holds the pad up with the notes facing Anne so that she can see them.

Therapist: Place your hand against this, against the writing on the pad.

Anne: *(holds the palm of her hand against the pad, fumbling a bit)* Okay...

Therapist: *(presses her hand against the pad from the other side)* Keep it in check. Keep it away from you!

Anne: *(puts more weight behind her pressure)* Like this?

The therapist and Anne press the notepad from their opposite positions, making the pad go back and forth a bit.

Therapist: Good! Remember what this is like *(lowers the notepad)*. Notice what that was like and compare it to this *(places the notepad in Anne's lap, with the scribbled notes facing up)*.

Anne: *(after a moment of silence)* Well, that's a little different.

Therapist: In what way?

Anne: Well, it's easier to hold it in my lap—in one way. But it's awful too. I don't like that.

Therapist: No. Which of the two positions requires more strength?

Anne: Well, that would be keeping it away from me. But that feels good too, in some way. At least I'm doing something.

Therapist: Okay. Which of these two positions is more similar to what you are doing out there, in your life?

Anne: Keeping it away from me. That's what I do all the time. That's it. That's what I do all the time, actually.

The dialogue during and after an exercise like this one can evolve in different directions. Sometimes clients remove the notepad as soon as it's placed in their lap. It's important for the therapist to be open to the client's actual experience and remember that this is what she should base her work on throughout. The dialogue is not an attempt at persuasion; it's a way of trying to evoke behavior that is clinically relevant and helping the client notice it. There are many possible aspects to map out, based on this exercise. What are the advantages of these different behaviors, and what are the disadvantages? In which of these two positions is the client more free to pursue a valued direction in life? If, like Anne, clients can clearly see that "keeping it away" is what they do, what would "keeping it in their lap" involve, if they were to do this in day-to-day life?

In these types of therapy situations, the goal is to map out and influence two types of behavior: problematic behavior in the shape of experiential avoidance, and alternative behavior. The exercise above, with the notepad, can suggest what the alternative is. Remember that, according to RFT, the fundamental problem with experiential avoidance is that verbal behavior in the shape of rule following is overly dominating. Exercises like the one described above can provide the client with a way of contacting an alternative, based on direct stimulus functions. For a short moment, verbal control is circumvented.

USING METAPHORS AS THERAPEUTIC TOOLS

Experiential exercises are used in therapy to try to dodge the dominance that indirect stimulus functions so easily acquire for us humans. Another tool that can be used for the same purposes is metaphors. Using metaphors as tools in psychological therapy is not a new device, and it definitely isn't a unique contribution from RFT and the research it is based on. This therapeutic tool has a long history within psychotherapy, especially within the nonempirical traditions (Evans, 1988; Kopp, 1995; Mays, 1990). At the same time, with the help of RFT, we can answer the question of why metaphors can be such an efficient tool. Understanding the significance of metaphors and analogies within human language also gives us some guidelines for how they can be used.

Metaphors are verbal tools. At the same time, they also contain relational networks where direct stimulus functions play a decisive role. Recall the metaphor that Sue and Larry are like cats and dogs, from chapter 5. Here are two relational networks. We have Sue and Larry on one side, and cats and dogs on the other. The relation between these networks is arbitrarily established through the contextual factor "are like." But each network also contains a number of direct stimulus functions that are established in the listener's history; for example, the experience of cats and dogs and how they usually relate to each other. And the point in this metaphor is that the relations between these direct stimulus functions can influence the listener's way of relating to Sue and Larry. In an instant, as soon as the metaphor is uttered, a large number of relations between stimuli are cued for the listener. These relations between stimuli are collected from events the listener has experienced and are then transferred to a phenomenon that is less well-known. If metaphors suggest possible alternative ways of acting, these are often flexible and open, rather than exact. This creates a shortcut that bypasses the verbal traps that could occur if the same behavior by Sue and Larry were to be described in more factual and didactic language.

In Anne's case, one way of talking with her about her attempts to solve her problems by thinking more could be saying, "What if you can't plow a field by tilling the soil in your head?" This metaphor is designed to help her contact the hopelessness of the strategies she's been using to try to solve her problems.

So what do we want to use metaphors to do? The same thing which we want to do by using a functional analysis. We want to help the client tact his own behavior, both the problematic behavior in the shape of experiential avoidance and alternative behavior.

Let's return to the dialogue with Anne. She actually offers a metaphor herself as she attempts to describe what she does when she thinks about her troubles and what she tries to do about them. She says that she "dives right in." This metaphor is a way of describing her particular variety of experiential avoidance. Perhaps it offers the therapist an inroad to discussing what Anne is experiencing, and maybe it also suggests a way out of her dilemma. The therapist could, for example, talk about how Anne experiences the trouble in so many different contexts and that she can't get away from it. It is as if life "throws her into the water." But even if she is thrown in the water, perhaps she can choose what to do once she finds herself there. For example, there may be a difference between "diving" and "swimming toward something you want to reach."

Although metaphors can often provide a shortcut that bypasses verbal traps, using them also entails some dangers. It isn't unusual for a metaphor to have certain functions for one individual and other functions for another. This is why it is important for the therapist to listen carefully to the client's experience. What doesn't work for the client doesn't work. Metaphors should not be used to argue with clients about what is right and what is wrong. If that is where you end up, therapist and client are probably in the grip of verbal traps once again.

The border between metaphors and experiential exercises, which can be used to evoke behavior, is often indistinct. This stands to reason, as it is the similarity between what is done in the first arena and what the client is used to doing in the second arena that is at the core of experiential exercises. Similarity of some kind is also the essential ingredient in a metaphor—a similarity between two relational networks.

Consider again the exercise involving the notepad, wherein the therapist tried to help Anne discriminate two different behaviors in relation to the pain she experiences: keeping it away from her versus keeping it in her lap. If, as in the dialogue above, the exercise leads to the client discriminating the difference between these two as relevant to actions out there in real life, then the exercise can be used as a type of metaphor. This can be done directly after the exercise via questions from the therapist: "If you were to keep these troubles (lack of concentration, tensions, troubling thoughts) more in your lap in your daily life, in what way would you do that, do you think?" The therapist could also return to this at a later stage, when the client relates something that has taken place in the second arena. The therapist may ask, "What you are describing now—would you say that was pushing it away from you or keeping it in your lap?"

Both experiential exercises and the use of metaphors are thus ways of trying to minimize the traps inherent in rule following and are therefore fundamental throughout therapy. This does not mean that rule-governed behavior is absent in the type of therapeutic processes illustrated in these examples. The client is a verbal being and will constantly create self-rules in these situations, independent of whether the therapist is taking an active part in this process or not. The hope is that these rules can become more flexible if they are based, firstly, on direct stimulus functions through experiential exercises and, secondly, on metaphors that make use of the client's direct experiences.

GOALS AS A PART OF THERAPY

It is obvious to most people that psychological therapy should have goals. In a functional perspective, goals are essential not only because they are what therapy is supposed to lead to; they are a part of therapy as such.

In therapy based on functional analysis, the decisive factor is the consequences of behavior. What governs behavior is first and foremost its past consequences. For verbal beings, another factor is what we could call desirable consequences. Through relational framing, we have the opportunity to seemingly contact events that have not yet taken place. We have the ability to want something we have not experienced, and we can act to make that thing happen. As described in chapter 6, this is the very basis of rule-governed behavior. This is the same behavior we refer to in ordinary language as having intentions: We act with a purpose.

The two types of clinically relevant behavior that I have described (problematic behavior and alternative behavior) are, of course, also this kind of behavior—behavior with the intention of achieving something; behavior with a certain direction. Behavior is influenced by desirable consequences. This was discussed in the dialogue with Anne, as the therapist inquired about the aim of Anne's behavior of "diving right in." Its aim was to dissolve the threat Anne experienced. It was aimed at getting rid of or gaining control of her tension, lack of concentration, tiredness, and thoughts of something being wrong with her. Perhaps Anne would say that she was striving to "feel normal." Anne's problem was that "diving right in" did not accomplish this and even blocked her from trying other solutions. Experiential avoidance dominated her behavior in general.

When trying to find an alternative behavior, it is therefore important to ask what the purpose is. If Anne is supposed to work on not diving right in by instead doing something else, what is the purpose of this? In which direction does she want to head? What are the desirable consequences she might work toward? If the therapist were to ask Anne, she might reply, "I just want to feel like I used to. I just want to be normal." But note that this leads to the same dilemma. These are precisely the type of goals that haven't worked for Anne. That this is the case is connected to the built-in problem with experiential avoidance: Thoughts, feelings, and bodily sensations usually aren't under voluntary control. Anne needs to direct herself toward something else, especially since refraining from diving right in and instead practicing doing something else could initially mean increased discomfort. After all, diving right in is a solution that's tied to a hope that it will help. If Anne refrains from this behavior, she might initially feel like she is worse off—as if she has given up. So what could motivate Anne to expose herself to this? She needs something that makes it worthwhile. Sometimes it might be enough to contact a possibility such as "My problem may get better in the long run." But clients often need something more significant, something they value—something that functions as a desirable consequence and that motivates them in experimenting with alternative behaviors. So figuring out the direction in which the client should turn is an important part of therapy.

In the dialogue with Richard, he describes that he "is just hanging in there" and agrees that he's "lying low." The therapist needs to work with Richard to define more specifically what this behavior consists of and what its purpose is. Richard hints that the purpose is preventing his problems from getting worse, because he doesn't want to experience what he previously experienced. He remembers an experience that he wants to avoid. But this direction could turn into a trap, as experiential avoidance so easily does. It will be important for Richard to identify what an alternative direction would be. If he were to refrain from lying low or just hanging in there, what would he do that for? What would the aim of his alternative behavior be? What would make it worth the risk? What would the desirable consequence be?

The attentive reader may notice that I am now approaching a certain type of rule-governed behavior: augmenting, which I discussed in chapter 6. In a model of psychotherapy based in RFT, there is an emphasis on the difficulties and traps inherent in rule-governed behavior. But there is more; such a model of psychotherapy also draws on the significant contribution of rule-governed behavior in creating more flexibility in human behavior. We can seemingly contact something we want and make that our direction, even when what we desire is not directly accessible. It need only be present to us through our ability to frame relationally. This can be very useful. Clarifying in which direction the client wants to move and the purpose of change is an important part of therapy. In everyday language, we might refer to this as motivation. I'll return to this in chapter 11, which focuses on working on rules, or verbal antecedents.

THE FOUNDATION OF A CLINICAL BEHAVIOR ANALYSIS THAT INCLUDES RFT

In a clinical functional analysis, we are interested in two types of clinically relevant behavior. One is problematic behavior or a behavioral excess belonging to a broad functional class: experiential avoidance. The other is an alternative behavior. In principle, the latter can be any form of behavior, as long as it doesn't belong to the class of experiential avoidance and aims for something else—preferably a consequence that is globally desirable to the client. The therapist's task is to help the client to reduce experiential avoidance and experiment with alternative behavior.

This may sound simple, but it rarely is. If it were simple, the client would probably have done it already. The side effects of human languaging block the way. Derived (indirect) stimulus functions contribute to the type of traps described in chapter 7 and summed up in the term "fusion." As humans, our actions are often fused with our own languaging; we follow explicit or implicit rules even when this type of behavior does not help us reach what we want. We are thus trying to achieve two things in therapy, and these primary therapeutic goals run through an individual's course of therapy as a whole, without necessarily occurring in any particular order.

- The therapist attempts to undermine nonfunctioning rule following and acts to help the client defuse, to shift to behavior that is more flexible in relation to indirect stimulus functions.

- The therapist tries to help the client change direction and turn toward things he values, things he wishes to accomplish or be true to. The therapist helps the client define consequences that are globally desirable for him and also helps the client act in a way that increases the probability of contacting those consequences.

Therapy thus takes into account the side effects of human languaging, as well as its possibilities.

All behavior is influenced by the context in which it occurs. It is influenced by the things that follow upon behavior (consequences) and by the things that precede it (antecedents). By altering these two aspects of the context, the therapist can influence the client. It is therefore natural to categorize therapeutic strategies accordingly, into strategies that primarily alter what follows upon behavior, on one hand, and strategies that alter what precedes behavior, on the other. Chapter 10 will focus on consequences, and chapter 11 will focus on antecedents.

CHAPTER 10

Altering the Context with a Focus on Consequences

When we say that the therapist, in her work with a client, can focus on either the consequences or the antecedents that surround a specific behavior, it is important to clarify one thing: We can never truly address only one or the other. All kinds of behavior have both antecedents and consequences that exert different types of influence on what takes place. In the type of interaction involved in clinical work, the therapist's actions will always provide both consequences and antecedents for the client's behavior. In the same way, the client's actions will provide both consequences and antecedents for the therapist's behavior. There is no escaping these conditions of human interaction. There is no neutral ground, no way of "doing nothing." If the therapist remains silent after the client says something, this will be the consequence the client faces after saying what he said. If the client then goes on to say something more, the therapist's silence was a part of the antecedents that preceded this action by the client. If the therapist says something, looks away, changes position, or acts in any other way, it is part of the antecedents and consequences that the client encounters in the moment and that influence his further actions. Mutual influence is a part of the human condition whenever two people interact. There is no other possibility. And this influence takes place through the consequences and antecedents established in the interaction.

When I write about focusing on consequences (or antecedents), I am saying that, in a given moment, the therapist can focus on one or the other. It is simply a way of describing the therapist's intention. If she has an intention of focusing her actions toward establishing consequences that reinforce a certain behavior on the part of the client, this is not a description of the entire process that occurs when she does this. In fact, there are inevitably many things occurring alongside whatever the therapist is trying to do, and the therapist is unaware of many of them. It would probably be impossible to describe everything that goes on. At best, the therapist is clearly aware of one or a few essential variables that she can alter.

INFLUENCING THROUGH CONSEQUENCES: THE CLASSICAL STRATEGY OF BEHAVIOR ANALYSIS

The most important conclusion in Skinner's work was that consequences of behavior are of primary significance in influencing subsequent behavior due to how different consequences reinforce or punish the behavior that precedes them. As a result, behavior analysis is strongly dominated by attempts to establish consequences that change behavior. Most of this work has been aimed at helping individuals with fairly serious problems, such as mental retardation and autism (Smith, 1996; Remington et al., 2007). In order to exert influence by establishing direct consequences for behavior, the therapist has to be present when the problematic behavior occurs. Since people with these types of problems have traditionally often been institutionalized, more possibilities for such influence have existed. Autistic children have been treated in accordance with these principles in their home environments, as well, though this requires the therapist to spend many hours a week with the child (Cohen, Amerine-Dickens, & Smith, 2006).

As pointed out earlier, this intense interaction between therapist and client doesn't occur within the framework of clinical settings. Psychoanalysis is the type of therapy that has the most time-consuming framework of interaction, but this too is sparse compared to the forms of therapy that applied behavior analysis has traditionally offered its highly functionally impaired clients. In regard to frequency of contact, clinical behavior analysis is more like other forms of psychotherapy. Therapist and client generally meet once or possibly twice a week. The total number of therapy sessions ranges from a few to many over the course of one or even several years. In this respect, clinical behavior analysis doesn't differ from what is standard in the types of psychological therapy normally called psychotherapy.

Considering the behavioral principles that form the foundation of behavior analysis, exerting influence by establishing direct consequences for behavior is a natural and fundamental approach. This also applies to clinical behavior analysis, although it has limitations because of the relatively sparse contacts between therapist and client. Let's take a look at how this influence can be exerted despite the time limitations.

ESTABLISHING CONSEQUENCES FOR PROBLEMATIC BEHAVIOR

In establishing consequences for problematic behavior, two clinically relevant classes of behavior are in focus: experiential avoidance in its various forms and alternative behavior. Let's return to Richard to illustrate how the therapist can establish consequences for Richard's problematic behavior in a way that can contribute to change.

Based on the dialogue in the preceding chapter, Richard and the therapist have worked together to identify Richard's passivity as a problematic behavior. He is lying low, standing by, and tagging along. That this is a problem for Richard is clear to him in one way, because he isn't satisfied with how things turn out. Together, they have also identified the reason for this behavior, as seen from Richard's perspective. As things stand, he behaves in this way so that things won't get worse. But he has also described this behavior as a longstanding pattern in his life. Just tagging along and accommodating himself to other people's wishes has felt natural to him; it feels like the safest strategy.

In conversation with his therapist, Richard has described what would be a desirable alternative behavior for him. He would like to be more assertive—take up more space in his interactions with other people. At work, this would mean letting his supervisor know what tasks and assignments he would like. At home, this would mean taking initiative in planning activities. For example, Richard would like to take the children to a football game. He's a longtime fan of the local team and has noticed that his children are interested in football. His wife, Erin, on the other hand, isn't very interested in sports. Richard and his therapist have discussed this as the basis for a homework assignment in therapy, that Richard should bring this up with his family. Afterward, the following dialogue takes place during session eight.

Therapist: How did things go with the football game? Did you bring it up?

Richard: Yes, but we just talked loosely about it. I checked what they thought about it. But there's a lot going on these next few weekends.

Therapist: The children have a lot planned?

Richard: Yes, Erin wants us to go to visit her sister. And next week is a busy week in school.

Therapist: How do you feel about this?

Richard: I'm not sure. You know how hard it is for me to pursue something that involves Erin. Can you help me out here? How do you handle women who always want to make all the decisions?

At one level, Richard is asking for advice. The therapist could suggest a solution and maybe do some role-playing around how Richard might act in relation to Erin. At the same time, the therapist notices that Richard is acting in his typically problematic way, here and now. He is lying low and seeking an initiative from the therapist. The therapist brings this fact into focus.

Therapist: I'm wondering a little about what you're doing right now, here, with me. I get the feeling that you're lying low—waiting to see if I'll take charge or give you something you can tag along with.

Richard: What am I supposed to do? You're the expert.

Let's stop and examine what the therapist is doing and what the result is. The therapist notices a problematic behavior on Richard's part: his typical version of experiential avoidance, what he does because it feels safest. Richard is lying low, taking a passive attitude, and seeking initiative from someone else, in this case the therapist. At this point, the therapist establishes a consequence that consists of simply tacting Richard's behavior, the very behavior they have defined together as problematic. Technically speaking, the therapist tries to establish a punishing contingency. She does something intended to make it less likely that Richard will repeat this behavior. We don't yet know what function this response will have for Richard's behavior, but we can assume that the therapist intends to establish an aversive consequence. How are we to understand this—that when she describes what Richard does using everyday language, she acts in a way that can function as a punishing contingency? We can understand this through RFT. In an earlier dialogue, this very behavior—withdrawing and lying low—has been defined as the problem. What Richard does increases his difficulties. When as a consequence of his actions Richard encounters this comment, it puts him in touch with "I am doing what gives me my problems." The therapist tacts Richard's behavior in that moment, and since this tact is in a relation of coordination to "what I do that makes things worse for me," it is likely to be aversive to him, provided that he actually shared the initial assessment that this behavior is something that contributes to the pain he experiences.

Although we assume that this action on the part of the therapist can make a positive contribution to change in Richard's life, it also involves two difficulties. The first is the terminology. Surely, no therapist wants to punish her client; this sounds inhumane. In order to avoid this unsympathetic terminology, therapists and theorists have sought other words to describe this process, and it might be wise to avoid a terminology that causes misunderstandings. Some speak of weakening a behavior instead of punishing it. Another way of putting it is to say that you block or obstruct the client's problematic behavior. The important thing is to understand what we are referring to, regardless of which terminology is used. The therapist establishes a consequence for the client's problematic behavior, and this is intended to reduce the chances of the behavior being repeated. In basic behavioral terminology, this process is called a punishing contingency, if it really has the intended effect.

The second problem with this course of action is more important than terminology or the risk of misunderstandings. When an aversive consequence is established, there is always the risk that avoidant behavior will simply increase in response. Indeed, this is exactly what happened in this example. Richard took an even more passive attitude: "You're the expert." Let's look at how the therapist handles this.

Therapist: Yes, well, that is my role. That's how it works. And of course this is exactly what we expect from experts: to give us something we can hang on to. You definitely have a point there. But you and I have talked about how this becomes a kind of a trap for you—that you tend to start lying low, just waiting for others to take the initiative and then you tag along. And even if this is what I'm supposed to do, as some kind of expert, I wonder if you might be getting stuck in this trap now, right here in our conversation?

Richard: *(first silent, then sighing)* Yeah, I can see what you're saying. I think that's right. I want someone to give me the answer. And that makes me even more passive. I start backing off. It's so hard.

Therapist: Yes, it is hard. And if you were to make some more room for yourself right here and now while we're talking—just when you feel that it's hard—how would you go about that? Is there a crack in the wall somewhere, a place where you could start to create some room?

Establishing punishing or weakening contingencies in relation to problematic behavior can be an important part of the work of focusing on consequences in the interaction between therapist and client. In this example the therapist retains the aversive consequence ("Now you are demonstrating your problematic behavior") while also doing a couple of other things in tandem. First, she says things that are likely to reinforce certain parts of Richard's behavior right now, the parts that are more functional. She can do this because there are several strands in Richard's behavior. While he does, in fact, lie low and demonstrate a passive attitude, he also continues the dialogue. At one level he tells his therapist some of his thoughts about the situatiuon. When the therapist confirms details he mentions, this hopefully has a reinforcing function in relation to these parts of his behavior. This is referred to as "validation" in dialectical behavior therapy, a therapy model that emphasizes that there is almost always something in what the client is saying that can be validated (Linehan, 1997). In more classical behavioral language, you might say that there is almost always some aspect of the client's behavior that should be reinforced. This is how the therapist views Richard's behavior when he says, "You're the expert." This is why she tries to establish a consequence to increase the probability that Richard will express his opinion again. The therapist tries to reinforce the functional behavioral class "expressing one's opinion."

The next thing the therapist does in the dialogue displays the impossibility of stringently separating the work connected to consequences from the work connected to antecedents. She establishes a verbal antecedent intended to function as a rule for alternative behavior. She inquires into the possibility of Richard himself taking up more space here and now—the possibility of Richard taking the initiative, so to speak. And she does this by referring to what she and Richard previously discussed as an alternative path: taking up space.

If Richard decides to take the risk to actually take up space, they could go on to talk about the difficulties involved in taking up space here and now, and about how it turned out. Then the dialogue could turn to how Richard might do this in relation to Erin. At that point, the intervention clearly would have moved on to dealing with antecedents for behavior—a behavior that is to take place in the second arena of this therapeutic work. But in this scenario, it would be done after Richard had more direct experience with both of the relevant behavioral classes. He would have demonstrated both the problematic behavior and the alternative behavior in the first arena and would have encountered different consequences for each.

ESTABLISHING CONSEQUENCES FOR ALTERNATIVE BEHAVIOR

When the therapist thinks she can discriminate the client doing something in the first arena that is an alternative to the problematic behavior, she should take the opportunity to establish a consequence aimed at increasing the probability of the client doing it again. In the dialogue above, the therapist did this in connection with Richard continuing the dialogue and expressing his opinion. This was done primarily to annul the risk of marked avoidance in the moment. Generally speaking, the behaviors we want to reinforce most of all are the ones that are the clearest examples of alternative behavior. Which behavior is, then, to be reinforced?

To answer this question, it's helpful for the therapist to keep in mind the definition of problematic behavior as experiential avoidance. The fundamental problem is the human tendency to use rule-governed behavior in an attempt to control, reduce, or get rid of private events; when it comes to feelings, memories, bodily sensations, or thoughts that spontaneously present themselves, we attempt a solution that often doesn't work. The alternative behavior to be reinforced is thus behavior that aims for something other than avoidance of private events. In principle, it could be essentially any behavior with such an aim. However, a great many of such behaviors probably don't aim for something the client desires. Therefore, we may need to delimit this further: The behavior to be reinforced should be behavior that has a higher probability of leading the client in a desired direction—with the exception of simply getting rid of private events.

Obviously, this solution contains a paradox. People seek psychological help in order to get rid of private events like depressive mood and anxiety. And successful psychological therapies do achieve this, to a greater or lesser extent. Yet this presents a dilemma: When getting rid of private events is the goal, behavior often ends up counteracting its own purpose, as described in chapter 7. This is an example of the remarkable side effects of verbal behavior.

Working with the client to map out what the alternative behavior should aim for is an important part of therapy. This is, again, an example of focusing on antecedents, since a current goal is a verbal statement (or thought) about a desired consequence, and not the consequence in itself. The consequence remains to be experienced. Richard has described what he wants to do as "taking up more space." When he was asked what he would do with this extra space or in what way taking up this space would be positive, he mentioned two things. First, he wants this to affect his relationships with other people. He wants to be closer to his children, to be more involved in their lives, and to share his own experiences. And at work he wants to prove his abilities and have enough space to do things he thinks will be beneficial for the students and the school. He doesn't like being passive and just tagging along. He wants to be a person who can take the initiative, at least more than he currently is.

In the functional analysis that the therapist and Richard have created together, it is clear that his passive attitude increases in situations where "things feel uncertain." Externally the situations are different, but they have in common the fact that other people are acting to achieve what they want or what they believe in and Richard notices that he wants something else. Even at this early stage, things begin to "feel difficult," so he waits. He feels that others disregard him, and he tends to ruminate over why it turned out this way and conclude that this is the inevitable outcome.

An alternative to lying low and waiting is, of course, to step forward and take up some space in the situations that feel difficult and uncertain. One situation where this behavior might be demonstrated in relation to the therapist is in regard to homework assignments. Let's assume that the therapist suggests a homework assignment in which Richard brings up with a colleague that he would like to switch tasks during an upcoming teachers' seminar. The therapist clearly shows, in various ways, that she thinks this is an appropriate homework assignment. If Richard objects and tells her that he would like to structure the homework differently, this could be seen as an instance of a desirable alternative behavior in relation to the therapist. This provides an opportunity to establish a reinforcing consequence, which the therapist could do by simply listening to Richard and taking his objection seriously.

Determining, in the moment, what therapist behavior will function as reinforcing for the client's alternative behavior is a difficult task. A supportive question the therapist can consistently ask herself is what behavior on her part might resemble a naturally occurring consequence in the client's day-to day life (the second arena) that would reinforce alternative behavior. In this light, the task in therapy is to offer a consequence in the current interaction that corresponds to this. In the example with Richard, his behavior of objecting and giving his own opinion can be reinforced by giving due consideration to what he says.

Another important question for the therapist to keep in mind is whether the alternative behavior is actually reinforced in their interaction. Is Richard's behavior of taking up space in his interaction with the therapist increasing as the therapy sessions proceed? If not, reinforcing contingencies haven't been sufficiently established, per definition. If this is the case, the therapist needs to examine her own behavior and probably needs to change it. If, on the other hand, Richard does demonstrate more alternative behavior in relation to the therapist, she can try to note the types of situations where this happens, and thereby gain a better understanding of which parts of her own behavior are benefiting the therapeutic goals.

It is important to note a risk involved in this work: that the reinforcing contingencies that are established may not be natural (Ferster, 1967), in which case they may not function as intended. Or they may function as reinforcing in interactions with the therapist, but not support generalization of the behavior in the client's life outside of therapy (Kohlenberg & Tsai, 1991; Tsai et al., 2008). It's especially important to be vigilant against establishing consequences aimed at influencing the client's behavior in a direction the therapist desires that doesn't align with the client's goals. If this occurs, it would be what in everyday speech we call manipulation.[1]

I described above how the therapist's tacting of the client's problematic behavior can function as an aversive consequence for this behavior. In a corresponding way, the therapist's tacting of alternative behavior can function as a reinforcing consequence for that behavior. Whether this happens depends on whether the functional analysis that defines a desirable alternative behavior is adequate. Let's suppose that Richard and the therapist agree that the behavior of "taking up more space" is desirable for Richard. And maybe, in his interactions with the therapist, this behavior has sometimes been followed by reinforcing consequences. When this behavior takes place again, perhaps the therapist says, "It just occurs to me that what you're doing now is different; that it's like you're making more room for yourself." As a result, in addition to the reinforcing consequence of obtaining something he wants, Richard now contacts the verbally established consequence "doing something new that helps me make changes." This is probably a reinforcing consequence for this behavior, provided that what the therapist says corresponds to Richard's own experience. If it does not, there is the obvious risk that the interaction instead becomes yet another example of Richard doing what he so often does: accommodating himself and tagging along. In this case, the therapist's comment would actually function as reinforcing for Richard's problematic behavior. This is another example of how important it is for therapists to be aware of the actual results of their interactions with clients.

---

1 Within behavior analysis (and experimental science more generally) the word "manipulation" is used without the negative associations it has in ordinary contexts. "To manipulate" simply means "to change something." In this sense of the word, when you do or say something in an interaction, it is always a manipulation of the factors that precede and/or follow the other person's behavior. The risk described here, however, is that the therapist's actions also become manipulation in the everyday, negative sense of the word.

EXTINCTION OF PROBLEMATIC RULE FOLLOWING

Based on RFT, we can draw the conclusion that rule following often contributes to psychological problems, as described in chapter 7. Establishing new direct contingencies in the interaction between client and therapist can contribute to breaking down this type of destructive verbal control. The direct experience of getting things you want through a behavior (for example, through "taking up space") can, in the long run, circumvent a behavior governed by a rule (for example, "I'd better lie low"). Of course, the dominance that verbally established contingencies often have for humans is not absolute. Behavior is constantly shaped by the interaction between the two essential paths to learning: through direct stimulus functions, and through derived (indirect) stimulus functions. New direct experiences that occur in the therapeutic interaction can break the dominance of verbal rules in a specific regard.

An analysis according to RFT provides yet another primary way of altering consequences in ways that can be helpful for the client. Let's return to the therapist's dialogue with Anne. During their fifth session they're talking about Anne's difficulties with concentrating and how this disrupts her work. Anne describes how, during a meeting at work, she realized that she hadn't caught what someone else just said. She became tense and thought, "Here I go again." Thoughts about how things would never work out started to come up, and she wondered if others noticed how weird she was feeling.

Anne: It's hopeless. I don't know what's going on.

Therapist: It sounds like this is difficult for you. And you want to get a grip on what's going on.

Anne: I can't live like this! Something has to be done about it.

Therapist: Okay then, there are several thoughts here: One is that you can't live like this. Another is that something has to be done about it.

Anne: I have to find a solution.

Therapist: So that's another thought: You have to find a solution.

Anne: What do you mean? I don't get it?

Therapist: Another one: I don't get it.

Anne: But... Are you saying that I'm just imagining this, or what? I really can't live like this. I have to be able to go to a meeting. Otherwise I won't be able to do my work. That's not okay. I have really tried, but it doesn't work.

Therapist: I hear another one: That's not okay.

Anne: You're making me confused. I'm really trying to understand, but I'm completely lost now...

Therapist: Good!

Anne: Good? In what way? What do you mean? I'm sure there is a point here, but I'm not getting it...

Therapist: You're trying to understand? Trying to work it out? What if it's like this: If you think you understand what I mean—whatever I mean, *that's* not it?

Anne: Should I get mad or start laughing? I'm totally confused...

Therapist: Exactly. Can you be there?

Anne: Confused? I don't like it.

Therapist: Okay. Can you have the experience that you are confused and don't like it, and just be there?

Notice how this conversation deviates from what we are used to in ordinary conversations. But the therapist has an explicit goal in her behavior of deviating. Languaging is a behavior that is maintained by a number of different contextual factors. It's a social game we play together and is dependent on a certain context. Once we've learned how to play, we can do so privately or silently to ourselves, but this behavior is originally learned in a social context. In a dialogue, our languaging is influenced by the consequences it encounters. In our everyday lives, we all inhabit a social and cultural context that offers a wealth of consequences that reinforce verbal behavior. People listen to what we say, and they answer in a way that we can understand. Even if they reject us, they do so in a way that follows the collective rules that are built into our languaging. Words usually mean what they normally mean. We usually adhere to the rules of grammar, and if we break these rules, we generally do so in a way that is "understandable." Even our breaches of the rules follow the rules of the game, so to speak. All of this is so natural that it's hard for us to imagine anything else. Now, if verbal behaviors are a part of psychological problems, the contexts that support these behaviors could be altered to help clients out of verbal traps. This is what Anne's therapist is trying to achieve in this dialogue. She does this while focusing on offering consequences other than those that normally maintain Anne's problematic verbal behavior.

Anne says that "something has to be done" about this. In an ordinary dialogue, listeners will, in one way or another, act on the indirect stimulus functions that this statement has in relation to them. They may, for example, suggest something that could be done. Or they may agree that something should be done, although they don't know what. Even if they reply by saying that it may not be necessary to do something about the situation, this is also a behavior that reinforces the verbal behavior in itself. It is coherent with verbal behavior as a whole, and as explained in chapter 4, coherence is a generalized reinforcer that is established early on in a child's language training. This is clear if Anne, in response to a potential reply that nothing needs to be done, gives reasons for why something actually must be done about this. In a case like this, Anne is performing even more of the verbal behavior she is so accustomed to performing. She's accustomed to thinking about what she should do, about why it doesn't turn out as she'd like, and about what's wrong. Also, just like all verbally competent humans, she is accustomed to giving reasons why she acts as she does. And the normal consequence of giving reasons is that the other party either accepts her reasons or lets her know why those reasons are not accurate or applicable. By not offering these types of reinforcing consequences for Anne's verbal behavior, the therapist is aiming at extinguishing Anne's problematic behavior within the therapeutic interaction.

First the therapist focuses on direct stimulus functions. Anne's behavior is dominated by the indirect stimulus functions of her statement. She acts in fusion with these functions; her behavior is dominated by them. The therapist does not, however, go along with this, as we normally do in a verbal dialogue, but instead points out what is actually present: "Look, here's a thought." When this is repeated Anne is confused and possibly frustrated. The normal way of speaking (and thinking) is not working. The consequences are not what they usually are. Anne first acts as though her description is being called into question, and she delivers reasons for why she has to do what she does. Yet the consequences still are not typical. She handles this as she is accustomed to doing: She increases her efforts to understand the situation. The therapist then says something that is likely to be aversive if you are really out to understand what's going on: "If you think you understand what I mean—whatever I mean, *that's* not it." How is such a statement to be handled?

The last comment, which becomes a self-conflicting rule for the behavior of trying to understand, could constitute a punishing contingency. This speaks to the risks associated with this type of contingency. As I pointed out earlier, it is important for the therapist to do something more than just obstruct problematic behavior. In this case, she brings in an antecedent for alternative behavior by hinting that maybe Anne can stay precisely where she is. A different way of making the suggestion would be a comment such as "What if confusion isn't your enemy?" For Anne, this could function as a rule for the behavior of staying where she is, feeling what she feels, and not being so quick to try to control or eliminate her uneasiness in the moment. The way I am here describing an intervention that establishes a prospective rule is another example of how working with consequences is never entirely about consequences, although consequences are what I have chosen to focus on in this chapter. It is also possible that the comment about confusion not being the enemy actually functions as a reinforcing consequence for parts of Anne's behavior. Even when Anne notices her own confusion and tacts this state in the moment, this behavior can already be different from the problematic behavior displayed at the beginning of the interaction. Staying in the moment and simply observing what there is to observe, without immediately looking to get away from her own state, is different from what Anne has typically been doing. In this way, the therapist's comment can further reinforce this rudiment of change.

RFT can inform and support this way of breaking the rules of languaging in order to help people get out of psychological traps. Similar strategies are a core component in acceptance and commitment therapy. It is important to understand that this is not something that should be done randomly in therapeutic work. It should be based on a functional analysis of an individual client's difficulties. In the example above, the therapist chooses to act this way in connection to Anne's typical problematic behavior of "diving right in" to a verbal way of solving a problem. Although Anne's strategy is intended to lead her to increased understanding and to answers to her questions, it ultimately leads to a dead-end street in her life.

So how is it possible for this new behavior to be generalized? After all, I have just explained that we constantly face a social context that, on the contrary, supports problematic verbal behavior. The answer lies in direct contingencies. It is common for clients to feel that the consequences of these types of interventions relieve some of the pressure they feel. When people who have struggled to do something that hasn't delivered on its promises finally realize that they can abandon this struggle, they often contact reinforcing consequences. An increased experience of vitality in the moment is common, as are opportunities for more flexible behavior. In this situation, people simply tend to do things they haven't done earlier. Anne may notice something that she previously just ignored. Perhaps she continues to interact with someone in a situation that she once would have cut short. Over time, this will lead to contact with a number of naturally occurring reinforcers that she hasn't contacted before. Perhaps she'll encounter a new type of response from someone. Perhaps she'll manage to carry out a work assignment that had frequently been interrupted as she switched her focus to how poorly she concentrates. Once Anne has experienced various new contingencies, something else occurs, as well. If Anne, in a certain situation, once again "dives right in" to her old, problematic ways of responding, it's more likely that she'll lose this new, rewarding experience. That could then function as a naturally occurring negative punishment contingency for her old way of responding.[2] This contingency can then also contribute to maintaining Anne's new behavior.

CONSEQUENCES AND THERAPY'S TWO ARENAS

In this chapter I have emphasized consequences for the client's behavior that are available during therapy sessions, what I've called therapy's first arena. These are consequences that the therapist can establish in a direct way. Throughout therapy, it is important for the therapist to be observant of the role she plays in the client's behavior and to make use of the opportunities for direct influence that might benefit the client. At the same time, it is in the second arena that clients live their lives, and this is where they ultimately need to change. In the end, the consequences the client encounters "out there" are the crucial ones, and these are out of the therapist's control. Still, the relationship between a behavior and its consequences is not one-sided. Just as the consequences of a behavior influence future behavior, the behavior influences what consequences will follow. If the client learns a new way of responding in interaction with the therapist and subsequently tries this behavior in the second arena, the chances of experiencing new consequences increase, and this can have a reinforcing function.

However, establishing consequences for current behavior is not the only way to influence future behavior. All behavior is also subject to influence by antecedents. Through arbitrarily applicable relational responding, antecedents become enormously more complex than if they were based only on direct contingencies. The possibility of exerting influence through verbal antecedents is fundamental to all models of psychotherapy. Skinner pointed this out

2 Naturally, this does not mean that anyone is punishing anyone else. I remind the reader of the definition of the term "punishing" within behavioral theory: a consequence that leads to a reduced probability that the behavior that preceded it will reoccur.

long ago, noting that giving advice and instructions are basic components in all kinds of therapy (Skinner, 1974, p. 204). Both the possibilities and the specific problems involved with verbal antecedents have been described in chapters 6 and 7. In the final chapter of this book, we'll return to this topic and its clinical implications.

CHAPTER 11

Altering the Context with a Focus on Antecedents

All psychotherapies take rule-governed behavior for granted; clients are assumed to have this capability. All therapists, regardless of psychological model, assume that instructions of one kind or another influence behavior. This is as true of the psychoanalyst who encourages her client to associate freely as it is of the cognitive therapist who asks her clients to note automatic thoughts or perform an experiment. It is also true of the therapeutic model described in this book. At the same time, an analysis of psychotherapy according to RFT involves an emphasis on the importance of understanding just how this process works.

Analyzing talk therapy in a scientific way requires a theory of language. The analysis of human languaging accounted for in part 2 of this book confirms the universal notion that verbal antecedents in the form of "talk" (aloud or silent) contribute to and influence other kinds of behavior. This provides us with further possibilities for influence, compared to behaviors that are governed purely by direct contingencies. However, as noted, this behavioral repertoire has some problematic side effects. A therapy based on this viewpoint should thus make use of verbal behavior while also remaining aware of its problematic aspects. Since RFT's insight into the problematic aspects of language is in many ways a novel contribution to the field of psychotherapy, this chapter will devote a fair amount of attention to the work that concerns the side effects of languaging. I will begin, however, with the more typical use of instructions as a means to achieving change.

VERBAL ANTECEDENTS FOR NEW BEHAVIOR

The therapist establishes verbal antecedents for new behavior as soon as she performs the initial functional analysis. She initiates rule-governed behavior in the client by asking the client to describe events that have taken place outside the consulting room; for example, things that happened earlier and the client's thoughts and feelings on these occasions. In behavioral terms, this is described as tacting (together with the client) clinically relevant behavior and its connections to contextual factors.

This is the actual basis of the therapy as a whole, and it is in itself an example of how, through verbal antecedents, the therapist can increase the probability of the client doing something different from what he usually does. The therapist may, for example, ask the client to fill out a log between sessions to provide information on the client's behavior in the second arena and what he does at different times of day. She will perhaps also ask him to make notes of his degree of depressive mood or anxiety in connection with the things he does (Dimidjian, Martell, Addis, & Herman-Dunn, 2008). If these directions cause the client to adapt and fill out a log as a homework assignment, this would be a typical example of rule-governed behavior. In addition, this is usually a new behavior. Clients (as well as people in general) do not usually take much time to observe, in such a structured way, the things they do and how this interacts with what takes place around them.

How such tacting of one's own behavior can in itself bring about changes has been discussed in chapter 9. An important part of the initial stages of therapy is for the therapist to reinforce precisely this type of self-discrimination. This is done by carefully monitoring the recording tasks assigned as homework. This is yet another example of how the therapist alternately focuses on influencing antecedents and consequences, respectively. Here we are focusing on antecedents, first and foremost, and more specifically on giving directions for how to carry out homework assignments like self-monitoring and recording of the client's own behavior.

In an initial therapeutic situation, such an analysis can quickly lead to a suggestion for alternative behavior in the second arena. Let's take a look at this dynamic in the example of Scott. For years, he's been experiencing that people talk about him, harass him, and want to harm him in different ways. This applies particularly to people he meets in a number of different everyday situations and who are unknown to him. All of this makes him feel unfairly treated and troubled, so he tries various strategies to avoid being exposed to these experiences. He avoids many settings where he feels he has a heightened risk of such experiences. As a result, he lives a rather isolated life and avoids social contact except with his family of origin. He lives alone in an apartment, and because of his difficulties, he is unable to work and is on disability. Sporadically, he hears voices that confirm his experiences. The voices speak maliciously of him and sometimes even threaten him. Most of the time he doesn't hear voices, but he still experiences the harassment going on "behind his back." Scott also does a lot of brooding on everyday events; he thinks about what others think of him, what they have said about him, and why.

In the initial functional analysis, the therapist and Scott identified two types of problematic behavior: He withdraws, and he also devotes a lot of thought to whether the events of each day are examples of harassment. These two behaviors seem connected in that the brooding is also a part of his strategy for not having to face perceived harassment. The experience of being someone who is harassed is painful; it brings his attention to his feelings of loneliness and how painful this is for him. In his fifth session, Scott relates an example of what typically happens.

Scott: I was standing in line, waiting for my turn at the counter. Then I hear some people in front of me say something I can't make out, and one of them looks at me. What the heck is that all about? I don't even know them. So what are they up to? My brother says they probably weren't talking about me, though...

Therapist: And what do you think?

Scott: *(sighs)* I don't know. Don't know for sure. But when I was standing in line it seemed very clear to me. That's exactly what happens—what always happens. Why do they want to pick on me?

Therapist: Okay, so whatever it's really about, we know one thing: You get this thought that they're talking about you. And that feels really lousy—like being harassed.

Scott: Exactly.

Therapist: So what do you do?

Scott: What do you mean? What can I do?

Therapist: I just mean, what did you do when this happened? You're standing there, you see them looking at you, you get the thought that they're talking about you, and you get the feeling of harassment. So what did you do?

Scott:	I just walked away to a different line. Tried not to look at them.
Therapist:	Do you think this could be an example of what we've talked about—that you withdrew?
Scott:	Yes, I think so. Yeah, that's what it is.
Therapist:	How did it turn out when you did this?
Scott:	Well, I got rid of them. But this is not how I want it to be—waiting in line all over again. Who wants that?
Therapist:	What would it be like, do you think, to get that feeling of harassment and simply stay in line?

This dialogue gives us examples of several fundamental behaviors on the part of the therapist. One is that the dialogue makes the therapist say something that could function as a rule for alternative behavior: "What would it be like to stay there?" At this point, of course, we don't know if this is going to become a rule for Scott. It is not until a behavior occurs in a certain relation to an antecedent that we can say the antecedent has functioned as a rule.[1] But it is possible that it will, and this possibility is one reason why the therapist says this. In such a dialogue, the therapist may, later in the session, propose a clearer suggestion (a rule) for Scott to use in a homework assignment by experimenting in a similar situation: first noting the "feeling of harassment," then staying there as an alternative to withdrawing. The purpose is to allow Scott to contact consequences other than those his current behavior results in. Please note, however, that this work is focused on antecedents; it is not the same as working to establish consequences for the problematic behavior being discussed. The therapist and Scott are talking about things that take place in the second arena. The only thing going on in the first arena, at the moment, is their talking. This means that the consequences that could follow

1 Let me remind the reader that the use of the word "rule" here deviates somewhat from the way this word is used in our everyday lives. Normally a rule is something with a specific expression. The definition is topographical. However, this is not the case here. Here a rule is an antecedent that has influenced a certain type of behavior: what we call rule-governed behavior. It is the behavior and its learning history that define what precedes the behavior (the antecedent) as a rule. Something that topographically would be defined as a question ("What would it be like to stay there?") can acquire the function of a rule.

upon the behavior they are talking about (withdrawing or staying there) are not consequences the therapist can control. They are indeed speaking about consequences, but that is quite different from establishing consequences. The only behavior that can be reinforced, or possibly punished, is behavior occurring in the moment. And when it comes to the behavior of carrying out a functional analysis, there is an impact. The therapist reinforces this behavior in different ways during the dialogue by rewarding Scott for describing events from the second arena and for his effort to discriminate different contingencies: what he does, on one hand, and what precedes and follows his behavior, on the other.

To first perform a functional analysis in this way, then reinforce this behavior in the therapy session, and, finally, based on the analysis, mutually arrive at alternative behavioral strategies is the core of the therapeutic model called behavioral activation (Martell et al., 2001; Dimidjian et al., 2008). These alternative strategies are then given as homework assignments for the client to experiment with between sessions. The experiments are subsequently analyzed, and new strategies are formulated and tested in new homework assignments. The aim of therapy is thus establishing of verbal antecedents—rules—for the client's actions in the second arena. The decisive question is, of course, whether the new behavior actually encounters reinforcing consequences. As mentioned, these consequences are out of the therapist's control.

In the same way that experiential avoidance is at the core of the analysis used to establish consequences, as discussed in the previous chapter, experiential avoidance is also at the core of the work to establish antecedents for alternative behavior. This is clear in the dialogue with Scott. The analysis accomplished in the client-therapist dialogue points to Scott's avoidance of "the feeling of harassment." The therapist tries to help Scott to tact the connection between the behavior and its consequences ("How did it turn out when you did this?") and introduces a possible antecedent for alternative behavior ("What would it be like to stay there?"). The basic strategy is simple, but doing this in a way that helps the client is often complicated. The task is to formulate possible rules in a way that influences the client to go on to test and try new behavior.

USING PLIANCE TO PRACTICE TRACKING

In the process outlined above, the type of rule-governed behavior aimed at is well-functioning tracking. This means the client follows a rule that states what is to be done in order to achieve a desirable consequence, and that the behavior works independently of the rule giver. When the behavior works, the client contacts the desirable consequence.

Let's say that Scott stays in the original line (or in a similar situation) while still feeling harassed, and the consequences seem positive to him in some respect. He may notice that it is easier for him to do his shopping, that the feeling of being harassed wanes, or that people don't actually talk about him in the way he had feared. Or maybe he discovers that they sometimes do, but he is able to interact with them in a way that works for him. If Scott keeps following the rule of "staying there" when his feelings and thoughts are occupied with the issue of being harassed, and he does this based on apparently contacting these new consequences through the rule, then the process is an example of tracking. If he continues to behave in a similar way several more times, the direct contingencies will take on a more dominant function. The direct experiences of what actually works will take over. Scott contacts new consequences because of his new behavior.

In many therapeutic situations, at the outset it can be difficult to get the client to experiment in this way. People who are depressed or who have very strong assumptions about the way the world is may not be ready to experiment to see what happens. Remember, too, that the consequences the therapist wants the client to contact are not something the therapist can control. Suggesting to Scott that he should experiment with "staying there" when he feels harassed doesn't guarantee that the consequence he will encounter if he does will actually reinforce his behavior of staying there. One way of tackling this potential difficulty is to begin with experiments where the therapist feels that reinforcing consequences are very likely to occur. Another alternative is to work with consequences that the therapist actually can control. For example, the therapist could say that certain homework assignments "are always included in this therapy" and make it clear that, in order to complete the therapy, the client must carry out the assignments. And in fact, this component is a constituent of most therapeutic setups, since adapting to the therapeutic model is a part of all therapies. If the client doesn't comply, he runs the risk that therapy will be suspended. Another example of controlling consequences would be if the therapist asked the client to do something for the therapist's sake. In both of these examples, the therapist is in command of the consequences specified in the rule.

If the client follows the rule "...because his therapist wants this" or "...because that's what you're supposed to do in this type of therapy," this behavior would be an example of pliance. While it is often problematic for pliance to dominate (as explained in chapter 7), attempts to establish pliance can nevertheless be used as a first step in an effort to establish well-functioning tracking. Something that is initially done "because the therapist wants it done" can, of course, also result in consequences beyond those the therapist is in control of, and in this way the behavior can become a first step in strengthening tracking. As described in chapter 6, this is the way we have all learned to track from the outset. Scott might initially "stay there" instead of walking away because he wants to satisfy or please his therapist. Even so, this may be an avenue toward encountering new, alternative consequences in the second arena. If these consequences are reinforcing for the behavior of "staying there," his behavior may gradually shift to tracking. The rule-governed behavior of staying rather than withdrawing—even in situations where he feels like people are talking about him—will then be working for him, independently of the rule giver.

Even if the therapist doesn't try to establish pliance, there is good reason to assume that this behavior plays a role in many therapeutic interactions. Clients often do things because they want to satisfy or please the therapist. While this can be a step in the process toward well-functioning tracking, if it blocks the path to change, it can be a problem. The latter aspect is something I will return to later in the chapter.

UNDERMINING VERBAL ANTECEDENTS

We all live under the assumption that it works to give instructions or rules, both in life in general and in psychological therapy. However, an analysis of verbal behavior according to RFT shows us that giving instructions does not always work as we assume it does, that it can be problematic, and why this is so. As described in chapter 7, human language has a built-in rigidity that makes us prone to different kinds of traps. We act on indirect stimulus functions even when doing so doesn't take us where we want to go. You might say our languaging has a fundamentalist quality and that this aspect is maintained by a number of different contextual factors that we are all collectively contributing to in this social game of languaging.

One essential way of altering these factors in the therapeutic interaction is by not offering the consequences we normally do following verbal behavior. This was described in the dialogue with Anne in chapter 10. Another way of altering the contextual factors in verbal behavior is to undermine verbal antecedents that influence problematic rule governing and establish alternative antecedents. Let's look at how Scott's therapist uses this technique.

Scott has told his therapist about his thoughts that others are talking about him and about his feeling of being harassed. He has given different examples of situations when this occurs. The following dialogue is from later in his fifth session.

Therapist: We've talked a bit about painful thoughts and feelings that you experience in different situations. I thought we might take some time to talk about thoughts and feelings in general, about how they function—things that are true about troubling thoughts but that actually apply to other kinds of thoughts as well, any kinds of thoughts and feelings. Would that be okay?

Scott: Sure.

Therapist: One thing is that they usually shift back and forth a bit. They come and go, so to speak. Take thoughts as an example. Do you notice having any thoughts right now?

Scott: Nothing particular...

Therapist: Let me give an example. Right now, when you said you didn't think of anything in particular, I thought, "What should I say next?" That was a thought that came up in my head.

Scott: Okay, I get it. So now I had a thought that I get the hang of what you mean. But that's the way it is all the time.

Therapist: Exactly. They come and go, and they shift back and forth. If you take a look around this room without doing anything else, I'm sure you will notice some more thoughts. Try it.

Scott: How? You want me to just look around and see if it makes me think of anything special?

Therapist: Yes, only it doesn't have to be something special—it can be anything. Take a couple of seconds to look around, and just make a note of the thoughts in your mind. You don't have to tell me what they are if you don't want to. Just try to notice them.

Scott: *(looks around silently first)* Well, first it was all just a blank, then I noticed thinking it's a blank. Then a bunch of other stuff; that I like that painting, for example *(points)*.

Therapist: Good! And isn't this the way it works most of the time? I mean, even when we talk to each other, there are simultaneously other thoughts in our heads—different thoughts than the ones we say out loud. At least that's how it works for me. I'm doing a lot of talking right now, but at the same time there are other thoughts running through my mind that I don't say out loud—thoughts that I'm the only one who notices. Are you with me here? Is that something you recognize?

Scott: Sure. There's no way you could make time to say everything.

Therapist: That's right. It's almost as if we could say there are four of us in the room. I'm over here, listening to my head, and you're over there, listening to yours. And all the time, we're talking to each other as well.

Scott: How do you mean, four? You're not talking about voices now, are you? I am not hearing any voices now.

Therapist: Well, I guess this may be true for voices too, but that's not what I meant. I mean this "talk" that's in all of us: thoughts, things we remember, things that come and go. But I guess it's the same thing with voices... If I get you right, they come and go too. Like you said, you aren't hearing any now.

Scott: Okay, I see what you mean. It sounds a little strange when you say it, but sure, that's what it's like for me too. Sure.

Therapist: And if you think about it, isn't that what feelings are like, too? They come and go. You feel one way right now; earlier today maybe you felt a different way. This morning I felt tired. Right now, though, I don't notice feeling tired at all. How do you feel right now, sitting here?

Scott: I'm fine. Right now, I'm almost having fun. But earlier, when I was waiting out there, things felt uncertain.

Therapist: That feeling of harassment, is that noticeable to you now?

Scott: *(after a moment of silence)* No, not right now. But when I first got here there was some of that. Not a lot, but a little bit. Out there in the reception room.

Therapist: Good, that's exactly what I meant. They come and go. Some thoughts or feelings are nice, others aren't that nice. Some we like, others we don't. And have you noticed something else? We're here the whole time, both of us.

	Thoughts and feelings come and go, but we, who notice them, we're here all the time.
Scott:	I don't know... I feel really insecure. I don't know who I am, really. It's like I'm sometimes this and sometimes that.
Therapist:	And when you feel like that, and you get the thought that you don't know who you are—who is noticing that at that moment?
Scott:	Noticing what it feels like, you mean? Well, I guess that's me. Yes, well, that's actually true *(laughs)*.

How are we to understand what the therapist is doing in this sequence? She tries to establish a context within which thoughts and feelings acquire a different function than they often do in our day-to-day life. She does this with a special focus on areas where we tend to get stuck in verbal traps. The traps that result in experiential avoidance come in different forms as described in chapter 7; however, all verbal traps are characterized by fusion, by behavior that is dominated by derived stimulus functions. You could say that we "take off" into the story offered to us by our own relating. Words and thoughts acquire power. This is what it's like for Scott. The combination of his life history and current circumstances provides him with his experience of being harassed. He thinks that others are talking about him, and he acts on these thoughts. One thing he does as a result is withdraw from people. Note that the crucial point in this discussion is not whether Scott's thoughts are what we normally call true. The point is that when he experiences these (true, untrue, or whatever) thoughts, he does things that clearly don't lead to good outcomes for him.

In the dialogue above, the therapist tries to establish a different context for these thoughts—a context that gives them a different function, even if they remain present. This is therefore an example of the therapist attempting to alter that part of the current context that serves as C_{func}. Her intention is to change Scott's behavior in connection with these thoughts, not to change or correct the actual thoughts. The therapist does not try to find alternative thoughts; she tries to help Scott develop an alternative behavior when these thoughts are present. Scott is used to behaving in a certain way based on the content of his thoughts. The therapist provides a context within which he can notice the process rather than respond to the content. She talks about thoughts (and feelings) the way we normally talk about external things: "Look, there it is." From the viewpoint of RFT, this means she is helping Scott take perspective on his private events. A verbally competent person always acts from the perspective of I-here-now. When we act in fusion with derived stimulus functions, these are also I-here-now. We find ourselves inside them. They establish our reality and thereby govern our behavior. As the dialogue above illustrates, the therapist can alter the context to shift what is felt and thought to I-there-then: "That was something I thought (or felt), there and then." The intention is not to discuss whether the thought or feeling is true or fitting; the intention is to help the client discriminate this in his own experience: "That thought (or feeling) is something that happens within me and, at the same time, at a certain distance from me." It is happening there. This isn't a logical argument. According to pure logic, our experiences do not occur at a distance from us. But as discussed in chapter 5, there is an experiential difference between our continuous experience of observing everything from a unique perspective ("I") and what we observe from this perspective. When our behavior is characterized by fusion, we frame these events within ourselves as I-here-now and not as I-there-then. In the dialogue above, the therapist is trying to help Scott develop this perspective-taking ability in relation to these private events.

This therapeutic strategy isn't intended as an argument in support of the therapist's views; it serves as a means of helping the client develop this type of self-discrimination. This makes it especially important for the therapist to be alert to the possibility that the client is simply agreeing with something she says, rather than discriminating his own experience. *This therapeutic strategy has a specific goal: to help clients frame private events as I-there-then. Private events that are framed relationally as I-there-then acquire different stimulus functions than if they were framed as I-here-now* (Luciano et al., 2004; Valdivia, Luciano, & Molina, 2006). Altering stimulus functions is the intention, because this leads to behavioral changes. The moment this perspective is taken (private events occur as I-there-then), the individual is in an experiential position that we might describe in everyday language as having a choice. In fused behavior, we don't experience such options; we simply act on the derived stimulus functions contacted, for example by "thinking." When the thought or the feeling is framed as I-there-then, we can still act on its content, but we can also act differently. This intervention is an example of what ACT refers to as defusion (Blackledge, 2007). As a therapeutic technique, defusion is synonymous with trying to break off fused behavior.

In the dialogue with Scott, the therapist established antecedents for behavior in relation to private events in general. Different types of thoughts and feelings are used as examples, and based on these, the therapist tries to help Scott to take perspective—to discriminate the private events in themselves as in some sense distinct from "himself." The thoughts and feelings that are the most difficult for him are mentioned, but they are not the focus. This can be an appropriate way of introducing defusion exercises in therapy.

We could see it as a form of didactics, serving to undermine normal antecedent functions and create a new context for private events in general. Later in therapy, however, it is important to follow up with defusion aimed at issues that are particularly relevant to the client's problem, as Scott's therapist does in the following example.

During his eighth session, Scott tells the therapist that he felt like people were looking at him in the reception room. He feels both upset and angry as he considers whether the people who sat there talking to each other were talking about him and why they would do that—what their reasons could be.

Therapist: So now you have the feeling of being harassed. And that's a difficult situation for you.

Scott: Why can't they just leave me alone? It's crazy! What do you think? How can they know things about me?

Therapist: It sounds to me like you're doing one of the things you usually do when you think and feel this way: You're trying to answer a whole lot of questions.

Scott: Yeah, that's true. I sure as heck don't want things to be like this.

Therapist: I wonder if we could do an exercise that might tell us something about what's happening right now, and that might be of help to you. Would that be okay?

Scott: Sure, sure—anything that helps me.

The therapist stands up and asks Scott to stand next to her. Then she picks up a newspaper and holds it close to both of their faces—closer than for normal reading.

Therapist: Are you able to read what it says?

Scott: No, not a chance. It's just a blur.

Therapist: Same thing with me. What if it wasn't just blurry, but scary, too, like the feeling of being harassed?

Scott: But in this case, you can back off or keep the paper at arm's length.

Therapist: You're right. Let's do that *(holds the paper farther away from their faces)*. How's that?

Scott: *(reads a few sentences from the paper)* Well, that's clear enough.

The therapist and Scott sit down.

Therapist: Maybe we could do something similar with the things you brought up—all of your thoughts concerning what happened in the reception room, your troubling thoughts and questions. Which ones are worst?

Scott: What do they want with me? Why don't they leave me alone?

Therapist: Are those the worst?

Scott: *(after a moment of silence)* I think the worst is this: They don't like me.

Therapist: Yeah, that is a troubling one. *(Silent for a moment.)* Do you think you could take that and keep it at some distance from you? Write it on the wall over there *(points to the wall behind them)*, below the painting. See if you can write it there, just so you can see it there in your mind's eye. You could close your eyes if that makes it easier.

Scott: That feels weird. I'd rather keep them open.

Therapist: That's just fine. See if you can write it on the wall: "They don't like me." Take your time, there's no hurry. *(She waits.)* Have you got it up there?

Scott: Yes. It's a little indistinct.

Therapist: That's okay; it doesn't have to be crystal clear. But you see it there?

Scott: Yes.

Therapist: Can you see what kind of handwriting you wrote in? You know, cursive or block letters, or what?

Scott: It's printed, with block letters.

Therapist: Good! Is it all in one line, or are the words below each other?

Here the therapist is applying the principle of defusion to the specific threatening content that Scott describes. The therapist helps Scott take perspective on this content with the intention of altering stimulus functions and creating a context where Scott can act more independently of the private events that normally dominate his behavior—those private events that were previously summarized in the expression "the feeling of harassment." The therapist uses the experiential distance that Scott has learned to contact between himself as perspective and the story that he is so easily absorbed in, the perspective-taking ability that is a result of ordinary language training. She also increases this experiential distance by helping Scott take the same attitude toward his experience of being harassed as he does toward external things. Notice, however, that the story as such is left as it is. The therapist even helps Scott phrase the most difficult part of the story before it is transferred from "here" to "there, on the wall." The follow-up questions about type of handwriting and so on are meant to function as antecedents for a diversified and flexible behavior in relation to that which is threatening.

This is a clear example of defusion in the shape of an exercise. But the same strategy can also be applied in typical therapeutic conversations. One example of this appeared in a dialogue with Anne in chapter 10, when the therapist repeats "That's another thought" several times. In that dialogue, the therapist's focus was on establishing consequences that could contribute to extinction of problematic verbal behavior. But such comments can also become antecedents for alternative behavior in relation to these thoughts. The therapist points out the possibility of regarding thoughts in perspective, as something there-then. Within ACT, this is an important strategy early on in the interaction with a client. When the client relates something that happened and his narrative includes a private event—for example, what he thought or felt—the therapist often replies by repeating what was related in a way that discriminates the feeling or thought in itself, relating to it as something there-then. If the client says that something isn't going to work, the therapist may reply, "So you had the thought that it won't work." As mentioned, the point is not to question what the client said. If the client sees the situation in that way and responds that "it definitely won't work," the therapist should clarify that the point in what she said wasn't to argue with whether a particular thought was correct. The point was to show that the statement or thought, in itself, is something that is uttered or thought. This is done with the intention of laying a foundation for more specific work on defusion by establishing antecedents for framing what was said or thought as I-there-then. This element is present in the first dialogue with Scott, when, after Scott has given his account of his painful experience, the therapist coins the expression "the feeling of harassment." In the parlance of psychology, we might say that certain thoughts, feelings, or other specific behaviors are, for a moment, acted upon as if they were objects. Notice, however, that this does not involve objectifying Scott. Quite the opposite. This is a conscious training of Scott's capacity to experience himself as the subject. For all of us, our experience of ourselves as subjects is based on precisely this type of training, in line with the processes outlined in chapter 5.

This discussion of undermining verbal antecedents is in no way intended to imply that verbal antecedents can be avoided in therapy; regardless of what the therapist says, it will have verbal functions for the client. But it is possible to alter the antecedents that typically influence problematic rule following. In practice, of course, this means that other antecedents will be established.

METAPHORS FOR ACHIEVING DEFUSION

Metaphors are another important tool for establishing antecedents that undermine problematic rule following and increase alternative behavior. A metaphor can suggest possible courses of action and thereby function as a rule, but it typically does so in a flexible and open-ended way, rather than by prescribing behavior precisely. Anyone who wants to follow directions conveyed via a metaphor will have to find his way forward step-by-step; his behavior will necessarily be characterized by trying it out. The overall direction is indicated, but no exact description of the behavior is available. This means the probability of contacting direct stimulus functions increases, which is exactly what the therapist is aiming for when striving to diminish dysfunctional rule following. To the extent that a metaphor contains relations between stimulus functions that have been established by the client's direct history, it will also have a guiding quality characterized by the client's experience.

A typical metaphor used within ACT to increase the probability of defusion involves passengers on a bus (S. C. Hayes et al., 1999). The therapist may bring this metaphor up in a situation where the I-here-now and I-there-then functions of private events have been brought to the fore. It can be used as a follow-up to an earlier defusion intervention, such as the previous dialogue with Scott, or as a part of an initial and more didactic phase of a therapy. It could be introduced as follows: "We could describe thoughts, memories, and feelings, and how they affect our lives, something like this: We are each driving around in a bus of our own. We are the ones doing the steering, and the bus is full of passengers: thoughts, memories, feelings, and so on. Passengers have been getting on at different occasions during our journey. Exactly when some passengers got on may or may not be clear. A simple example for my own part is that because I grew up in Chicago, I have images of Chicago as passengers on my bus. These passengers frequently take up space in the back of the bus, even though I don't go to Chicago very often nowadays. How about you—do you have any images of Chicago as passengers on your bus?"

It would be fitting to engage in a short dialogue to check that the client follows what is being described here. If he does, the therapist can go on to describe thoughts and feelings as other passengers on the bus. She can mention that some passengers seem rather neutral, others seem nice, and some seem unpleasant, perhaps even repugnant. She can bring in private events that the client has told her about and that seem to have a painful governing function in his life. For example, Scott's therapist could mention the "feeling of harassment" as one of the unpleasant passengers. The therapist can also talk about how we don't notice all of the passengers all of the time, but that in certain situations they stand up, or they may even come forward to breathe down our necks and tell us how to drive. In Scott's case, his therapist could continue by saying something like "Take that feeling of harassment, for example. It keeps coming up to the front of the bus, with lots to say, doesn't it? Pretty heavy-duty stuff, too. Would you say it's telling you how to drive, as well?"

Up to this point, the metaphor has established private events as being at a certain distance from the driver. Another aspect of the metaphor is how it can be used to talk about who is steering the bus. The therapist can talk about how the driver could drive according to what the passengers tell him or how he can choose to drive in spite of what they're saying. She can draw out the traps that the client (and people in general) sometimes ends up in by letting different passengers decide where to steer the bus; in Scott's case, "the feeling of harassment" makes him steer away from the line he had been waiting in. The therapist can talk about what we humans typically do to make passengers sit down quietly and not come forward to threaten or distract us. She can talk about how, even when hostile passengers are right behind us, breathing down our necks, we can still keep driving this bus of ours. No one else is at the wheel. There is a fundamental difference between the driver and the passengers.

This approach allows the therapist to speak metaphorically about experiential avoidance. At the same time, the metaphor can be used to help clients accept whatever they find threatening in the back of the bus, which will help increase their flexibility in relation to what they usually avoid.

ACT offers a number of metaphors and encourages therapists to make use of them (S. C. Hayes et al., 1999; Luoma et al., 2007). It also encourages therapists to develop new metaphors together with their clients—metaphors that are personally relevant and useful for particular clients. Of course, the therapist's ability to identify suitable metaphors requires a solid understanding of the principles of treatment in order to determine what needs to be highlighted. One metaphor for defusion that I use now and then is from a context most of us recognize.[2] It is a short sequence in the 1937 animated Disney film *Snow White and the Seven Dwarfs,* and in particular the scene where Snow White is in the cabin with the seven dwarfs, dancing to the music. Some of the dwarfs are playing musical instruments when, suddenly, a bumblebee flies in and starts to distract them. The dwarfs begin to adjust their behavior in an effort to get rid of it. Thoughts can be like that bumblebee—and maybe like a whole swarm of them. It's natural to do things like waving your arms to get rid of bumblebees. The experience of having a whole swarm of bees around your head can be rather threatening. But can we drive them away by desperately waving our arms? And what about the thoughts and recollections that are at the center of the client's pain and struggle? Can the client drive them away by making them the focus of his actions? What does the client typically do that resembles waving off bumblebees, and how does that usually turn out? This metaphor provides a means of talking about these experiences and about how the client acts when they pop up. But not only is this done in the frame of a metaphor; the very structure of the metaphor places the private events "out there," like bumblebees. In this way, their stimulus functions are transformed. In addition, the client's behavior can become less dominated by this experience, since a verbal being's actions are always necessarily performed from the position of I-here-now.

PRIVATE EVENTS AS OBSTACLES TO WHAT MATTERS IN LIFE

The clinical applications of RFT come in many varieties, and yet this work is, in many ways, just in its infancy. On one hand, RFT sheds new light on therapeutic strategies that are well-known in themselves, and on the other, it offers a number of new ideas and approaches. In this book, I focus on the interventions I consider to be the most groundbreaking: defusion and valued living that embraces painful private events. These strategies are also particularly emphasized within ACT. Yet not even these ideas are completely new. In various ways, they are also built into therapeutic models based on different theoretical backgrounds. The theoretical understanding that RFT offers can help define and improve these strategies further. This is natural, since RFT

2 This may be particularly true in Sweden, as this scene is a part of the Disney potpourri of classics and specials that has been aired on Swedish television every Christmas for the last fifty years or so. Thank you to the client who first brought this metaphor up!

is a theory that is concerned precisely with languaging as such, and all talk therapies primarily make use of language. From a deeper understanding of languaging, we are able to demonstrate in greater detail how these strategies should be developed and used.

The first strategy, described above, is contained in the term "defusion." Its object is to establish an increased distance between the client's experience of a continuous self (self-as-perspective, discussed in chapter 5) and the private events that have taken on dysfunctional control of the client's life. The goal, of course, is to thereby reduce the probability of experiential avoidance. The second main strategy is to work with the experiential obstacles the client experiences and their relationship to what the client feels is important in life (Luciano et al., 2004).

One of the basic principles of learning theory is that we humans are governed by consequences. The consequences of our past actions have a strong impact on our future behavior. This is the fundamental thesis in operant psychology. Once we have learned arbitrarily applicable relational responding, this capability changes what can function as governing consequences. As a result, external things and internal experiences no longer function only as what they are in themselves; they can also acquire a number of different functions depending on what they are related to. And since this relating is, to a high degree, governed by a socially arbitrary game, the variability is immense.

For nonverbal beings, the things that can function as reinforcing consequences are sharply delimited. Primary reinforcers can have such a function, and so can reinforcers that have been established by being directly connected with primary reinforcers. The same is true for stimuli that have a formal, physical similarity to any of these. For a verbal being, however, variability is increased because networks[3] of different derived relations can make an event reinforcing. For example, a consequence of a certain action can be reinforcing if it is in a relation of coordination with "being a good person." Yet it can also be punishing if it is in a causal (if-then) relation to "the future misfortune of my children."

3 At the risk of repeating myself, I want to remind the reader that this does not mean that such networks actually exist, either in the external context or in an imaginable internal world. This is simply a way of describing a behavior that is governed by complex derived relations. These relations in themselves are, in turn, a way in which the individual responds, governed by contextual factors. The reason I am repeating myself a bit in this regard is that our everyday language is so permeated with mentalistic concepts that they are hard to avoid, even when one strives to use a more stringent, scientific language.

These types of relational networks can also be an element of antecedents and thus function as rules and influence behavior. This is the very essence of rule-governed behavior. As defined in chapter 6, the type of rules that influence the degree to which different events function as reinforcing or punishing are augmentals, and the behavior that indicates that an antecedent has had an augmenting function is called augmenting. This is the phenomenon we are referring to when, in everyday language, we talk about what is important to us—the things we want and wish. More scientifically, these wants can be described as verbally constructed desirable consequences that are globally valid for a given individual. Within ACT, the word "values" is used, but the exact wording isn't important. We could just as well talk about what we want in life and what is important to a given individual.

The things that are important—the things we want our lives to be about—have a crucial function in our work for change. These are the things that motivate us. In order for us to want to make changes, there must be something at stake, some point in making different choices from those we have made before. We need to be able to see something we want ahead of us—a goal or valued direction. And when we do, we are in apparent contact with desirable consequences. These contextual factors can, in turn, function as antecedents for actions that increase the probability that we will contact the desirable consequences more directly.

People who are stuck in experiential avoidance have problems in this area. It isn't that they lack values, or things that make them want to act in a certain direction; the problem is that private events, in the form of thoughts, feelings, memories, or bodily sensations, have been established as obstacles to actions that could lead to these desired consequences. For these individuals, private events function as barriers to these types of actions. From their point of view, removing or controlling these private phenomena is a prerequisite for engaging in actions that move them toward their valued consequences.

To illustrate, let's return to Richard. He describes desirable consequences, consequences that he values. In the dialogue with his therapist, he has described what kind of a father he wants to be and how he would like to act at work. However, he refrains from actions that could lead in this direction, due to feelings of insecurity, memories of things that happened earlier when he took similar steps, and various thoughts associated with all of this. In short, for Richard these private events stand in opposition to actions that point in the direction he wants to go. This brings us to a more precise formulation of the second fundamental strategy for diminishing the control of verbal antecedents where they have problematic influence on the individual's life: *Alter the relations between the relevant private events and the valued actions so that rather than being in a relation of opposition, they enter into a relation of coordination.* In this case, a relation of coordination means that these two—the private events and the valued actions—can be present at the same time; they can coexist. Once again it is important to point out that this strategy is not about arguments or efforts to convince the client that "this is the way things actually are." The aim is to provide a context in the interaction between therapist and client that increases the probability of behavior where these two are in coordination.

Let's return to the two metaphors described above and look at how they can be used in this strategy. In the conversation about the metaphor of passengers on a bus, the therapist could begin by talking about values and the important things in life. "If you were free to drive anywhere you want, where would you drive your bus? In a world where you were totally free, what direction would you start driving in? What would you be heading toward in different areas of your life?" The therapist could also talk about the relation of opposition that is implied in the client's behavior and results in the client focusing his actions on "getting rid of the passengers"; for example, "It seems like you feel you need to get rid of your passengers first. What do you think would happen if you simply let them be and brought them along on your trip to where you want to go?"

The metaphor from the scene in *Snow White and the Seven Dwarfs* can be used in the same way: "What if, instead of waving your arms to make the bumblebees go away, you were to keep your hands on your instrument and play the tune you want to play? It takes guts, because it means you have to let the bees fly around wherever they want. But what if they aren't your enemies? What if they're not what they appear to be? What if you can just let them be there and still direct yourself toward what you aspire to?"

Another example of the same strategy is the exercise the therapist did with Anne in chapter 9—the one where the notepad was used to illustrate the choice of the two functional behavioral classes of "keeping it away from me" and "keeping it in my lap." The first choice keeps that which is threatening in a relation of opposition and establishes the experience as an obstacle that must be removed or kept away. The second allows those same threatening phenomena to exist in a relation of coordination with actions that Anne values. This helps her see that she can let her difficult experiences be there while simultaneously moving in the direction she wishes.

Note how a relation of opposition between private events and valued actions tends to lead to rigid behaviors and vicious circles. For Richard, a feeling of insecurity is in opposition to moving in the direction of things he wants. If, based on this relation, he refrains from doing what he wants to do, then the feeling of insecurity will become increasingly aversive over time. From the beginning it was a feeling that in itself was unpleasant, but with time it also becomes "what obstructs me from what I desire." And the more important this desire is, the more aversive the feeling of insecurity will be, precisely because it seems to block moving ahead in that direction. This makes it all the more important to get rid of, and so the cycle continues.

In efforts to eliminate experiential avoidance, these two strategies work together well. The first strategy, defusion, can undermine the tendency of private events to influence behavior here-now. This is done by framing the private events with problematic stimulus functions as I-there-then. Transforming their stimulus functions in this way makes it easier to relate to them as something that can be "brought along for the trip," instead of something that is in opposition to engaging in valued actions. As a result, they are also likely to become less aversive as time passes. Richard may still feel insecure in certain situations. But if he has brought this feeling along while acting in the direction of what he desires and experienced positive consequences of these actions, his relationship to his feelings of insecurity will change as well. They may still be unpleasant, but they are no longer aversive in the sense that they block his path toward what he sees as important in life.

NOT KNOWING WHAT ONE WANTS

In many cases it is easy to define therapeutic goals. People know what they want, what matters to them, and into which areas they want to tread. This is true even if they feel that there are obstacles to those goals. In situations like these, conversations about the client's goals and values function as motivational antecedents. With Richard, such a conversation could revolve around the role he would like in his work as a teacher or around how he and his children could spend their time together, perhaps by going to watch their home team in a game. The therapist speaks in a way that gives Richard the opportunity to "scent" something he would like to attain. If this contributes to Richard actually taking steps in that direction, what the therapist said may have worked as a motivative augmental (discussed in chapter 6).

But sometimes the situation in therapy is quite different. Nothing seems to be motivating for the client. The therapist's attempts to talk with the client about what would be worth reaching for go unanswered. The following dialogue, which takes place during Richard's ninth session, illustrates this situation.

Richard: Things keep going wrong. I feel like I'm trying, but it doesn't lead anywhere. Yesterday I actually suggested that we could go to the game tomorrow. But it's a night game, and Erin thought it was too late because my son has a major Spanish test the next day. It's the same way at work. I've tried to talk to the principal, but nothing happens. And when it comes to my work, I'm not sure I even know what I want. I think I've done my best, but...

Therapist: So you've come to a halt. Do you have any idea what you could do in this situation?

Richard: No, it seems hopeless. And I'm not really sure what the point is any more. Is work really that important? I don't know.

Therapist: What is important, then?

Richard: I wish I knew. That's exactly it. Well, my family is important. But they're doing okay either way. It's like it doesn't really matter. I don't really know what is important.

Therapist: That sounds like a good reason to just lie low. How does that sound?

Richard: No, that's no good. It's not good the way it is, I know that. I just don't know what I want. There's a void. I think everyone is so used to me lying low. That's how it has always been. It might be too late, I guess.

Therapist: And how does that sound; if it's too late?

Richard: *(noticeably affected)* Heavy. Too damn rotten. It's like I'm never given the chance. There's no space for me.

Therapist: *(silent for a moment)* You know what it makes me think when you describe these difficult things that are so heavy for you?

Richard: No, what?

Therapist: It makes me think that what is important, what matters to you, is right here. Right here, along with the heavy and the rotten stuff.

Richard: How do you mean? To me it's all just hell.

Therapist: I'm thinking of what you said about never getting the chance. That there's no space for you. There's something about this getting a chance—what it would be like if there was a space for you. It sounds important...

It's one thing to say that nothing is important, nothing matters, and you don't know what you want. That's verbal behavior, and pretty common, at that. And if a client says this, there is an important point in it. Many times it tells the therapist something about the difficulty the client finds himself in. But saying that nothing matters is rarely an actual description of the client's experience. In other words, it is rarely a tact and particularly not a pure tact. It is more often a mand. It is a verbal statement governed by earlier consequences of saying this, consequences like what the listener says or does at that point. One common consequence when someone says he doesn't know, or that nothing matters, is that the person he's talking with counters this somehow; the other person provides advice. Another common consequence is that the other person stops asking. The latter is especially common after someone says, "I don't know." Both of these consequences would probably lead Richard deeper into his typical dilemma.

The therapist does not, however, interpret what Richard says as a statement about what is at hand. The reason for this is that she doesn't think people are generally indifferent. As humans, we are governed by consequences. What follows upon our actions is of importance to us, just as is true for other animals. We can talk in different ways and for different reasons, but that doesn't change our basic conditions. In order to enter into a dialogue about this, the therapist draws nearer to Richard's pain and points out that what is important is in relation to what is painful. This isn't peculiar, when you think about it. In fact, it seems natural. Psychological pain exists precisely because something important is inaccessible. If nothing were important, why would anything feel difficult, rotten, or heavy? Richard's use of words also points to this link when he talks about not being given a chance and there not being space for him.

There are many fruitful ways of continuing this dialogue with Richard. Hopefully he will be able to see that there is something important here, in being given a chance and a space. The therapist can ask him how he thinks he'd like to use a chance or space if it were given to him. They can also talk about what would be necessary for him to take a chance himself or to take up space if this isn't "given" to him. The latter issue is probably important to discuss with someone like Richard. It's also likely that there will be reasons to return to the two strategies discussed above: first, to help Richard take perspective on the things that are heavy and painful, and second, to alter the relation of opposition between these experiences and actions that are directed toward taking up some space.

The point above, that there are always things that are important to us, no matter what we say, may need to be clarified. Even if uttering "I don't know what I want" often represents a mand and a quest for specific consequences, the difficulties in knowing what one wants can also be real. Among other things, knowing what one wants requires the ability to discriminate a number of one's own subtle processes. To learn this is to learn to tact one's own private events as they take place, developing what was called self-as-process in chapter 5. Some of us have had good training in this skill, and others have not. Some people have primarily learned to identify other people's actions and needs and therefore have developed generalized pliance. As a result, they may not be skilled at tacting their own private events. Another possible reason why an individual may have problems accessing what he wants is that certain private events may seem very threatening. This can lead to deliberately avoiding focusing on those events. Over time, this can have a side effect of making it difficult for the person to access other private stimuli, as well, beyond those initially avoided.

Learning to tact self-as-process can thus be an important skill in identifying what matters in one's life and therefore is also important in becoming motivated to choose one's direction. The work connected with this in clinical behavior analysis is similar to what has been described in other therapeutic models (Greenberg & Pavio, 1997). Taking it in small steps, the therapist works on helping the client to focus on and to discriminate what goes on in himself—usually in the form of emotional and physical phenomena. Alongside this, it generally remains important to work on defusion and clarifying values. This is because even the behavior of tacting what is taking place within oneself can become complicated in connection with rule-governed behavior. If this occurs, defusion exercises may be important to help the client develop more self-discrimination.

The issue of knowing what one wants gives us yet another example that the work with antecedents and the work with consequences always go together. The specific intention here, though, has been to point out the importance of talking about the important things in life as a way of establishing motivational antecedents. The therapist tries to establish conditions that function as augmentals for alternative behavior.

UNDERMINING DIFFERENT FOUNDATIONS IN THE FUNDAMENTALISM OF LANGUAGING

In chapter 7, I described different types of problematic rule-governed behavior. One common example of this is when pliance is generalized and thereby blocks well-functioning tracking. Another example is when tracking of fairly short-term consequences blocks the tracking of more long-term consequences.

The rigidity of this type of rule following is often increased by more comprehensive, abstract rules about what is important in life that strengthen the governing force of the consequences specified in the more concrete rules. In this case, problematic augmenting is dominant.

One important type of rule-governed behavior occurs when verbally constructed consequences that are globally desirable for the individual are governing. In the discussion above, I described how these types of consequences ("what I want," "what is important in life") can be used to motivate alternative behavior. But these rules can also participate in the client's problematic behavior. If we take a closer look at the three clients I've described, we can see that their problematic rule-governed behavior follows some rules that are rather concrete. However, these rules acquire their motivational strength from more comprehensive rules. The interventions I've described as ways of undermining verbal antecedents may need to address both the concrete rules and those that are more comprehensive and abstract.

Let's return to Anne. Her problematic rule-governed behavior (experiential avoidance) quickly becomes obvious in the dialogue. Bodily sensations and phenomena she doesn't recognize in herself lead to a self-generated rule about how the situation has to be solved, and these experiences are followed by behavior that is aimed at achieving this—among other things, what she metaphorically calls "diving right in." This behavior turns into an example of tracking where fairly short-term consequences of "feeling better" dominate and also block Anne's contact with more long-term consequences. Defusion strategies would involve undermining the governing antecedents by helping Anne take perspective on both the bodily sensations and the catastrophic thoughts she has in connection with them. The therapist would also work to change the functions of the physical sensations from being obstacles (in opposition to valued action) to being something Anne could bring along while she does what she wants to do.

At the same time the therapist could look for more comprehensive rules that may be at work. This may not be necessary, but the more difficulty Anne has in doing something other than what she normally does, the greater the need to seek motivational antecedents that exert strong control on her problematic behavior.[4] One way of doing this is by asking what makes it so important to her not to experience those physical sensations. Anne might reply that it simply "feels wrong," and that this indicates that something is wrong. Again, the therapist can try to help Anne to frame this as I-there-then rather

4 Of course, when carrying out a functional analysis, there is also reason to look for consequences that contribute to this situation.

than as I-here-now and to act not as though this is an obstacle, but instead as though it is something that can be brought along. In this type of dialogue it is fairly common for clients to say things like "It occurs to me that this is probably what I've always done—even before that terrible episode. I think it has always been important to me to have things feel normal. Nothing is supposed to feel disturbing." Many times this is followed by recollections from earlier events in life. If Anne were to say something along these lines, then based on RFT the therapist might suspect that this suggests a motivational antecedent to Anne's problematic behavior. It could be an augmental that makes it even more important to Anne to "dive right in" or otherwise do whatever she can to eliminate or control her private events. The same therapeutic strategies are still applicable in this case.

Theoretically, we are once again approaching a problem I briefly touched upon in chapter 6 in connection with the discussion about implicit rules. In the current discussion of rule-governed behavior, where is the rule? In this case it seems like a post hoc construction. Anne describes her behavior as if she followed a rule that nothing is supposed to feel disturbing. She may say that she often thinks in this way, which makes it clearer that we are dealing with verbal behavior here. A common variation would be demonstrated if Anne said that she hasn't thought along these lines for many years, but that she can remember often thinking in this way as a child, or that her parents often spoke in this way. But there is also the possibility that she doesn't recognize these thoughts at all, from recent times or the distant past. She simply notices her own behavior as something that she has done for a long time, that this is something she has always done. In this case, the question of whether this behavior should be regarded as rule-governed or as controlled by direct contingencies cannot be answered. We wouldn't know enough about Anne's learning history to determine this. Be that as it may, it is now possible for Anne to relate verbally to her own behavior. She can discriminate it as "something I do (there-then)" and thereby arrive at the experience of having a choice. At this point the therapist may ask a question such as "What would be important to you—what would you like to do if you weren't focusing on making sure nothing disturbs you?" Hopefully this would function as an antecedent for alternative behavior.

Readers who are accustomed to working within a psychodynamic or cognitive frame of reference will, of course, recognize these processes. We are now dealing with phenomena that in psychodynamic therapy fall under the heading of "the subconscious" and within cognitive therapies are associated with "schemas" or "underlying beliefs." These different theoretical constructs attempt to describe the same or similar phenomena. From a behavior analytic perspective, however, it is important to emphasize that we do not gain anything, in practice, by referring to unobservable internal structures in order to explain phenomena. Our focus is not on the rule, but on the behavior. The rule may be observable if it is uttered or thought by someone, but it can also be implicit and only exist as an abstraction of the interaction in the moment. In this case, the rule exists in the interaction between the behaving human and the conditions under which he acts. Nevertheless, it can be useful to talk about a rule in the therapeutic situation, as this can facilitate the client's taking perspective on his own behavior, which is the aim.

We can apply a similar analysis to the therapist's attempts to undermine different foundations in the fundamentalism of languaging in connection with Richard and Scott. I'll do so briefly to provide further illustration of these principles, starting with Richard. Richard describes a specific rule about lying low in relation to others who express what they want and think about things. The obstacle he runs into is his experience of uncertainty. This experience can probably be clarified and more closely defined in the ongoing client-therapist dialogue. The function of this experience (uncertainty) as an obstacle can be undermined using the same strategies as described for Anne. Richard's problematic behavior is clearly interpersonal and may be a sign of problematic pliance. If this is the case, the therapist has strong reasons to pay especially close attention to the interaction in the room, as mentioned in chapter 10. This can help Richard discriminate his own problematic behavior in the interaction, and it can also help him take perspective on his experiential obstacles and move in the direction he desires.

What would problematic augmenting look like in Richard's case? It could be displayed in the form of avoiding taking up space, based on an implicit (or imagined) consequence that involves being rejected or attacked—being seen as someone who isn't good enough. This rule can also be undermined through defusion. Is it possible for Richard to have this experience of being seen as not good enough, take perspective on this, and, having this experience (there-then), still act toward whatever he chooses or values?

The description of Scott's interactions with the therapist demonstrated the work to undermine verbal rules. Scott's behavior is governed by concrete rules about how to act when he feels harassed, and also by a more comprehensive rule that specifies the risk of not being well liked. The way Scott talks about this intimates that the therapist is closing in on something that for Scott (and for most of us) is a dreaded consequence: being abandoned and alone. A factual consequence like this is likely to be aversive to humans, so it's natural that we try to avoid it; however, when avoiding even apparently contacting this consequence in the form of a verbal antecedent, we tend to do things that don't work for our best interests and that often obstruct moving in desired directions. Helping Scott discriminate this verbal antecedent as something that he (here) can have (there), and helping him take this experience along while he turns toward the things that are of value to him, is the aim of the work in therapy. This is done in the hope that he will thereby encounter consequences that are better for him than those he is contacting by doing what he does now.

SUMMARY

There is room for more than one way of working with verbal antecedents within clinical behavior analysis. Giving instructions for alternative behavior plays an important role. This takes place explicitly when the therapist suggests something, but it also occurs implicitly when the client draws conclusions from what goes on in the therapy setting and then generates rules himself, based on this. Our knowledge of how this happens has been considerably increased through the research RFT is based on. My specific focus, however, has been on how we can work with the dark side of rule following by undermining the control that verbal antecedents have over human behavior. One particularly important factor in this is the function that private events acquire when we act to control or eliminate them. I've described two main strategies intended to alter the functions of private events that serve as problematic antecedents for behavior. One strategy involves developing the client's experience of a continuous self (self-as-perspective) and distinguishing the governing private events from that ongoing, continuous sense of self. When we humans act in fusion with private events, they serve as natural starting points and therefore dominate our behavior. When we frame them as I-there-then, their stimulus functions are transformed and we have a better chance of behaving flexibly in relation to these events. This is defusion, and it gives us a better possibility of successfully applying the other primary strategy I've described. The second strategy is aimed at undermining the function of private events as obstacles to actions that might take the client toward long-term and generally desirable consequences. The intention is to move from behavior governed by these experiences—particularly behavior based on these experiences being in opposition to valued actions—to behavior where these experiences are brought along in the direction the client wants to go. In both of these strategies, the therapist works to change the context within which the client's problematic verbal behavior occurs.

The focus on these two strategies is a direct outgrowth of an understanding of verbal behavior based in relational frame theory. This doesn't mean that these approaches are entirely novel or that there is nothing similar in other models of psychotherapy. However, the special emphasis on these strategies is new, and some concrete and innovative therapeutic moves follow from this.

One fundamental difference compared to most other models is taking behavior analysis as a starting point. There is good reason to return to this as the book draws to a close. The essential assumption is that behavior is performed by the organism as a whole, and the factors that govern this behavior are to be found in the context in which the behavior occurs. This applies to all behavior, including verbal behavior, which is often described in terms of language and cognition.

While this fundamental basis is limiting in some ways, it also opens the door to many potential interventions. It is limiting in that only the contextual factors are of direct interest. The therapist is a part of the client's context, and her only way of intervening directly is by altering that same context. At the same time, this starting point offers a great deal of hope. It is possible to alter contextual factors; they are accessible to direct influence. This provides the therapist with an opportunity to assist the client in making the changes he desires.

Afterword

There are at least two good ways to be practical. One is to tell someone how to do something in a given situation. That is a great way to make a difference quickly, but it does not tell the person what to do when the situation is different. Another approach is to tell a person how things work, and then to make suggestions about what to do based on that understanding. This approach is admittedly more complicated, but if it is successful, the person understands a good deal more. The principles are known.

A person with that level of understanding might see entirely different alternatives for what should be tried. New situations are less frightening precisely because they are a bit less new when their depths are plumbed. It is easier to be creative while still staying linked to what is known.

This book is a practical book of the second kind. Those of you who have stuck with the book now have a sense of what I mean. It is not by accident. Törneke had a purpose in mind: "to show how learning theory, including more recent findings in this field, is directly applicable in clinical practice" (p. 170). He wanted to describe "a psychological treatment that is based on principles rather than well-defined manuals" (p. 171). I believe he has done so, from the ground up. As supplemented by his book *The ABCs of Human Behavior* (Ramnerö & Törneke, 2008), this book carries behavioral thinking into a practical psychology adequate to serve as a foundation for clinical work, regardless of label or tradition.

This is a variant of an old idea—as old as behavior therapy itself. In the early days of behavior therapy (in the early 1960s), the idea was that behavioral principles could be scaled into a complete psychology of human struggle and achievement. Unfortunately, it was not to be. There were bold efforts, but the principles were not adequate to deal fully with human cognition. The vision foundered and practitioners turned to clinical and commonsense theories of cognition.

As Törneke has shown, it is different now. The bottom-up, inductive vision in behavior analysis has continued to unfold. There is now a lot more to be said.

There is nothing of importance in human cognition that is obviously outside of the theoretical reach of RFT. Empirically, much more remains, of course, and I do not mean to imply that RFT ideas are correct in any ultimate sense. All ideas are limited, and to some degree, that means all scientific ideas and theories are incorrect. There is no reason to suspect that has changed in the last day or month or year or decade. But even limited ideas can be useful for a time—perhaps a long time. Thus, I do not mean that this analysis is "right." I mean only that behavioral psychology is in the game. The analysis is plausible, empirical, and evolving. And unlike many theories of cognition that are out there, this approach tells practitioners what language and cognition is, how it comes to be, what its benefits and weaknesses are, and how it can be modified to better serve human interests.

The principles in this book give every sign of being useful for a time—maybe even a long time. Behavioral and RFT principles are not designed to apply merely to acceptance and commitment therapy, or even functional analytical psychotherapy or behavioral activation. They are designed to have broad scope and to undergird all forms of psychotherapy. They are designed to relate to all forms of education. These principles are designed to be applicable at home, in business, in the consulting room, or at school. They arguably apply to all verbal events, even the ones occurring at this moment.

I understand this sounds grand. It can't be helped. As Törneke points out, learning theory is inherently grand in its vision.

What readers can do with a book like this is not to "believe" but to step forward boldly and responsibly and apply their new understanding to see how well it works. What will your world look like clinically (and at home, at work, and in your relationships) as you begin to think about language, thought, and reasoning from a deeply functional and contextualistic viewpoint?

Answering that question is an exercise that is both fun and practical. It promises to liberate practitioners. For those with scientific leanings, it points toward a very different way to be empirically supported: know the principles, analyze the situation in terms of these principles, act based in part on that understanding, and evaluate how it turns out, recycling as needed. Instead of being restricted to empirically supported packages, we can explore empirically supported procedures, processes, and principles. That is a liberating vision worthy of effort.

Oh, and one final step. Go to www.contextualpsychology.com and let us all know what you find out.

—Steven C. Hayes
Foundation Professor of Psychology
University of Nevada

References

Abramowitz, J. S., Tolin, D. F., & Street, G. P. (2001). Paradoxical effects of thought suppression: A meta-analysis of controlled studies. *Clinical Psychology Review, 21,* 683-703.

Alexander, F., French, T. M., & the Institute for Psychoanalysis. (1946). *Psychoanalytic therapy: Principles and application.* New York: Ronald Press.

Andersson, J. R. (2005). *Cognitive psychology and its implications* (6th ed.). New York: Worth Publishers.

Ayllon, T., Layman, D., & Kandel, H. J. (1975). A behavioral-educational alternative to drug control of hyperactive children. *Journal of Applied Behavior Analysis, 8,* 137-146.

Bach, P., & Moran, D. (2008). *ACT in practice: Case conceptualization in acceptance and commitment therapy.* Oakland, CA: New Harbinger.

Barclay, M. W. (1997). The metaphoric foundation of literal language: Toward a theory of the reification and meaning of psychological constructs. *Theory and Psychology, 7,* 355-372.

Barlow, D. H. (2002). *Anxiety and its disorders: The nature and treatment of anxiety and panic* (2nd ed.). New York: Guilford Press.

Barnes, D., Hegarty, N., & Smeets, P. (1997). Relating equivalence relations to equivalence relations: A relational framing model of complex human functioning. *Analysis of Verbal Behavior, 14,* 1-27.

Barnes-Holmes, D., Barnes-Holmes, Y., & Cullinan, V. (2000). Relational frame theory and Skinner's *Verbal Behavior*: A possible synthesis. *Behavior Analyst, 26,* 69-84.

Barnes-Holmes, D., Barnes-Holmes, Y., Power, P., Hayden, E., Milne, R., & Stewart, I. (2006). Do you really know what you believe? *Irish Psychologist, 32*, 169-177.

Barnes-Holmes, D., Hayes, S. C., Dymond, S., & O'Hora, D. (2001). Multiple stimulus relations and the transformation of stimulus functions. In S. C. Hayes, D. Barnes-Holmes, & B. Roche (Eds.), *Relational frame theory: A post-Skinnerian account of human language and cognition* (pp. 51-72). New York: Kluwer Academic/Plenum.

Barnes-Holmes, D., Staunton, C., Barnes-Holmes, Y., Whelan, R., Stewart, I., Commins, S., et al. (2004). Interfacing relational frame theory with cognitive neuroscience: Semantic priming, the implicit association test, and event related potentials. *International Journal of Psychology and Psychological Treatment, 4*, 215-240.

Barnes-Holmes, D., & Stewart, I. (2004). Relational frame theory and analogical reasoning: Empirical investigations. *International Journal of Psychology and Psychological Therapy, 4*, 241-262.

Barnes-Holmes, Y., Barnes-Holmes, D., & McHugh, L. D. (2004). Teaching derived relational responding to young children. *Journal of Early and Intensive Behavior Intervention, 1*, 3-26.

Barnes-Holmes, Y., Barnes-Holmes, D., & Smeets, P. M. (2004). Establishing relational responding in accordance with opposition as generalized operant behavior in young children. *International Journal of Psychology and Psychological Therapy, 4*, 559-586.

Barnes-Holmes, Y., Barnes-Holmes, D., Smeets, P. M., & Luciano, C. (2004). The derived transfer of mood functions through equivalence relations. *Psychological Record, 54*, 95-114.

Barnes-Holmes, Y., Barnes-Holmes, D., Smeets, P. M., Strand, P., & Friman, P. (2004). Establishing relational responding in accordance with more-than and less-than as generalized operant behavior in young children. *International Journal of Psychology and Psychological Therapy, 4*, 531-558.

Barnes-Holmes, Y., McHugh, L., & Barnes-Holmes, D. (2004). Perspective-taking and theory of mind: A relational frame account. *Behavior Analyst Today, 5*, 15-25.

Baron-Cohen, S., Tager-Flusberg, H., & Cohen, D. J. (2000). *Understanding other minds: Perspectives from developmental cognitive neuroscience* (2nd ed.). New York: Oxford University Press.

Beck, A. T. (1963). Thinking and depression: Idiosyncratic content and cognitive distortions. *Archives of General Psychiatry, 9*, 324-333.

Beck, A. T. (1964). Thinking and depression: Theory and therapy. *Archives of General Psychiatry, 10*, 561-571.

Beck, A. T. (1967). *Depression: Clinical, experimental, and theoretical aspects.* New York: Harper and Row.

Beck, A. T. (1976). *Cognitive therapy and the emotional disorders.* New York: Meridian.

Beck, A. T. (1991). Cognitive therapy: A 30-year perspective. *American Psychologist, 46*, 368-375.

Beck, A. T. (1993). Cognitive therapy: Past, present, and future. *Journal of Consulting and Clinical Psychology, 61*, 194-198.

Beck, A. T. (2005). The current state of cognitive therapy. A 40-year perspective. *Archives of General Psychiatry, 62*, 953-959.

Berens, N. M., & Hayes, S. C. (2007). Arbitrarily applicable comparative relations: Experimental evidence for a relational operant. *Journal of Applied Behavior Analysis, 40*, 45-71.

Blackledge, J. T. (2007). Disrupting verbal processes: Cognitive defusion in acceptance and commitment therapy and other mindfulness-based psychotherapies. *Psychological Record, 57*, 555-576.

Blackledge, J. T., Moran, D. J., & Ellis, A. (2008). Bridging the divide: Linking basic science to applied psychotherapeutic interventions—A relational frame theory account of cognitive disputation in rational emotive behavior therapy. *Journal of Rational Emotive Cognitive Behavior Therapy, 27*, 232-248.

Bowlby, J. (1988). *A secure base: Parent-child attachment and healthy human development.* New York: Basic Books.

Buber, M. (1970). *I and thou.* New York: Touchstone.

Callaghan, G. L., Gregg, J. A., Marx, B. P., Kohlenberg, B. S., & Gifford, E. (2004). FACT: The utility of an integration of functional analytical psychotherapy and acceptance and commitment therapy to alleviate human suffering. *Psychotherapy: Theory, Research, Practice, Training, 41*, 195-207.

Campbell-Sills, L., Barlow, D. H., Brown, T. A., & Hofmann, S. G. (2006). Effects of suppression and acceptance on emotional responses of individuals with anxiety and mood disorders. *Behaviour Research and Therapy, 44*, 1251-1263.

Carpentier, F., Smeets, P. M., & Barnes-Holmes, D. (2002). Matching functionally-same relations: Implications for equivalence-equivalence as a model for analogical reasoning. *Psychological Record, 52,* 351-312.

Carpentier, F., Smeets, P. M., Barnes-Holmes, D., & Stewart, I. (2004). Matching derived functionally same relations: Equivalence-equivalence and classical analogies. *Psychological Record, 54,* 255-273.

Catania, A. C. (2007). *Learning* (4th ed.). New York: Sloan Publishing.

Chapman, A. L., Gratz, K. L., & Brown, M. Z. (2006). Solving the puzzle of deliberate self-harm: The experiential avoidance model. *Behaviour Research and Therapy, 44,* 371-394.

Chawla, N., & Ostafin, B. (2007). Experiential avoidance as a functional dimensional approach to psychopathology: An empirical review. *Journal of Clinical Psychology, 63,* 871-890.

Chomsky, N. (1959). Reviews. *Verbal Behavior,* by B. F. Skinner. *Language, 35,* 26-58.

Chomsky, N. (2006). *Language and mind* (3rd ed.). Cambridge, UK: Cambridge University Press.

Clark, D. M., Ehlers, A., Hackmann, A., McManus, F., Fennell, M., Grey, N., et al. (2006). Cognitive therapy versus exposure and applied relaxation in social phobia: A randomized controlled trial. *Journal of Consulting and Clinical Psychology, 74,* 568-578.

Cohen, H., Amerine-Dickens, M., & Smith, T. (2006). Early intensive behavioural treatment: Replication of the UCLA model in a community setting. *Developmental and Behavioral Pediatrics, 27,* 145-155.

Collins, A., & Burstein, M. (1989). Afterword: A framework for a theory of comparison and mapping. In S. Vosniadou & A. Ortony (Eds.), *Similarity and analogical reasoning* (pp. 546-565). Cambridge, UK: Cambridge University Press.

Dahl, J., Plumb, J., Stewart, I., & Lundgren, T. (2009). *The art and science of valuing in psychotherapy.* Oakland, CA: New Harbinger.

Devany, J. M., Hayes, S. C., & Nelson, R. O. (1986). Equivalence class formation in language-able and language-disabled children. *Journal of the Experimental Analysis of Behavior, 46,* 243-257.

Dickins, D. W., Singh, K. D., Roberts, N., Burns, P., Downes, J., Jimmieson, P., et al. (2001). An fMRI study of stimulus equivalence. *Neuroreport, 12,* 2-7.

Dimidjian, S., Hollon, S. D., Dobson, K. S., Schmaling, K. B., Kohlenberg, R. J., Addis, M., et al. (2006). Randomized trial of behavioral activation, cognitive therapy, and antidepressant medication in the acute treatment of adults with major depression. *Journal of Consulting and Clinical Psychology, 74,* 658-670.

Dimidjian, S., Martell, C. R., Addis, M. E., & Herman-Dunn, R. (2008). Behavioral activation for depression. In D. H. Barlow (Ed.), *Clinical handbook of psychological disorders* (4th ed.; pp. 328-364). New York: Guilford Press.

Dobson, K. S. (1989). A meta-analysis of the efficacy of cognitive therapy for depression. *Journal of Consulting and Clinical Psychology, 57,* 414-419.

Dollard, J., & Miller, N. E. (1950). *Personality and psychotherapy.* New York: McGraw-Hill.

Dougher, M. J., Augustson, E. M., Markham, M. R., Greenway, D. E., & Wulfert, E. (1994). The transfer of respondent eliciting and extinction functions through stimulus equivalence classes. *Journal of the Experimental Analysis of Behavior, 62,* 331-351.

Dougher, M. J., & Hackbert, L. (2000). Establishing operations, cognition, and emotion. *Behavior Analyst, 23,* 11-24.

Dougher, M. J., Hamilton, D., Fink, B., & Harrington, J. (2007). Transformation of the discriminative and eliciting functions of generalized relational stimuli. *Journal of the Experimental Analysis of Behavior, 88,* 179-197.

Dugdale, N., & Lowe, F. L. (2000). Testing for symmetry in the conditional discrimination of language-trained chimpanzees. *Journal of the Experimental Analysis of Behavior, 73,* 5-22.

Dymond, S., & Barnes, D. (1994). A transfer of self-discrimination response functions through equivalence relations. *Journal of the Experimental Analysis of Behavior, 62,* 251-267.

Dymond, S., & Barnes, D. (1995). A transfer of self-discrimination response functions in accordance with the arbitrary relations of sameness, more-than, and less-than. *Journal of the Experimental Analysis of Behavior, 64,* 163-184.

Dymond, S., & Barnes, D. (1996). A transformation of self-discrimination response functions in accordance with the arbitrarily applicable relations of sameness and opposition. *Psychological Record, 46,* 271-300.

Edvardson, C. (1997). *Burned child seeks the fire.* Boston: Beacon Press.

Ekman, P. (1992). An argument for basic emotions. *Cognition and Emotion, 6,* 169-200.

Ellis, A. (1989). The history of cognition in psychotherapy. In A. Freeman, K. M. Simon, L. E. Beutler, & H. Arkowitz (Eds.), *Comprehensive handbook of cognitive therapy* (pp. 5-19). New York: Plenum.

Ellis, A. (2005). Can rational-emotive behavior therapy (REBT) and acceptance and commitment therapy resolve their differences and be integrated? *Journal of Rational-Emotive and Cognitive-Behavior Therapy, 23,* 153-168.

Evans, M. B. (1988). The role of metaphor in psychotherapy and personality change: A theoretical reformulation. *Psychotherapy, 25,* 543-551.

Farmer, R. F., & Chapman, A. L. (2008). *Behavioral interventions in cognitive behavior therapy.* Washington, DC: American Psychological Association.

Ferster, C. B. (1967). Arbitrary and natural reinforcement. *Psychological Record, 17,* 341-347.

Ferster, C. B. (1972). An experimental analysis of clinical phenomena. *Psychological Record, 22,* 1-16.

Ferster, C. B. (1973). A functional analysis of depression. *American Psychologist, 28,* 857-870.

Freeman, A., Pretzer, J., Flemming, B., & Simon, K. (2004). *Clinical applications of cognitive therapy* (2nd ed.). New York: Kluwer Academic/Plenum.

Gentner, D. (1989). The mechanism of analogical learning. In S. Vosniadou & A. Ortony (Eds.), *Similarity and analogical reasoning* (pp. 199-241). Cambridge, UK: Cambridge University Press.

Gentner, D., Bowdle, B., Wolff, P., & Boronat, C. (2001). Metaphor is like analogy. In D. Gentner, K. J. Holyoak, & B. N. Kokinov (Eds.), *The analogical mind: Perspectives from cognitive science* (pp. 199-254). Cambridge, MA: MIT Press.

Gentner, D., Holyoak, K. J., & Kokinov, B. N. (2001). *The analogical mind: Perspectives from cognitive science.* Cambridge, MA: MIT Press.

Gifford, E. V., & Hayes, S. C. (1998). Functional contextualism: A pragmatic philosophy for behavioral science. In W. O'Donohue & R. Kitchener (Eds.), *Handbook of behaviorism* (pp. 285-327). San Diego, CA: Academic Press.

Gortner, E. T., Gollan, J. K., Dobson, K. S., & Jacobson, N. S. (1998). Cognitive-behavioral treatment for depression: Relapse prevention. *Journal of Consulting and Clinical Psychology, 66,* 377-384.

Greenberg, L. G., & Pavio, S. C. (1997). *Working with emotions in psychotherapy.* New York: Guilford Press.

Harmon, K., Strong, R., & Pasnak, R. (1982). Relational responses in tests of transposition with rhesus monkeys. *Learning and Motivation, 13,* 495-504.

Hayes, L. J. (1992). The psychological present. *Behavior Analyst, 15,* 139-145.

Hayes, S. C. (1984). Making sense of spirituality. *Behaviorism, 12,* 99-110.

Hayes, S. C. (1989). Nonhumans have not yet shown stimulus equivalence. *Journal of the Experimental Analysis of Behavior, 51,* 385-392.

Hayes, S. C. (1991). A relational control theory of stimulus equivalence. In L. J. Hayes & P. N. Chase (Eds.), *Dialogues on verbal behavior* (pp. 19-40). Reno, NV: Context Press.

Hayes, S. C. (2008). Climbing our hills: A beginning conversation on the comparison of acceptance and commitment therapy and traditional cognitive behavioral therapy. *Clinical Psychology: Science and Practice, 5,* 286-295.

Hayes, S. C., Barnes-Holmes, D., & Roche, B. (Eds.). (2001). *Relational Frame Theory: A Post-Skinnerian Account of Human Language and Cognition.* New York: Kluwer Academic/Plenum.

Hayes, S. C., Brownstein, A. J., Zettle, R. D., Rosenfarb, I., & Korn, Z. (1986). Rule-governed behavior and sensitivity to changing consequences of responding. *Journal of the Experimental Analysis of Behavior, 45,* 237-256.

Hayes, S. C., Fox, E., Gifford, E. V., Wilson, K. G., Barnes-Holmes, D., & Healy, O. (2001). Derived relational responding as learned behavior. In S. C. Hayes, D. Barnes-Holmes, & B. Roche (Eds.), *Relational frame theory: A post-Skinnerian account of human language and cognition* (pp. 21-50). New York: Kluwer Academic/Plenum.

Hayes, S. C., Gifford, E. V., & Hayes, G. J. (1998). Moral behavior and the development of verbal regulation. *Behavior Analyst, 21,* 253-279.

Hayes, S. C., & Gregg, J. (2000). Functional contextualism and the self. In C. Muran (Ed.), *Self-relations in the psychotherapy process* (pp. 291-307). Washington, DC: American Psychological Association.

Hayes, S. C., Kohlenberg, B. K., & Hayes, L. J. (1991). The transfer of specific and general consequential functions through simple and conditional equivalence classes. *Journal of the Experimental Analysis of Behavior, 56,* 119-137.

Hayes, S. C., & Strosahl, K. D. (Eds.). (2004). *A practical guide to acceptance and commitment therapy.* New York: Springer-Verlag.

Hayes, S. C., Strosahl, K. D., Bunting, K., Twohig, M., & Wilson, K. G. (2004). What is acceptance and commitment therapy? In S. C. Hayes & K. D.

Strosahl (Eds.), *A practical guide to acceptance and commitment therapy* (pp. 3-29). New York: Springer-Verlag.

Hayes, S. C., Strosahl, K., & Wilson, K. G. (1999). *Acceptance and commitment therapy: An experiential approach to behavioral change.* New York: Guilford Press.

Hayes, S. C., Wilson, K. G., Gifford, E. V., Follette, V. M., & Strosahl, K. (1996). Experiential avoidance and behavioral disorders: A functional dimensional approach to diagnosis and treatment. *Journal of Consulting and Clinical Psychology, 64,* 1152-1168.

Hayes, S. C., Zettle, R. D., & Rosenfarb, I. (1989). Rule following. In S. C. Hayes (Ed.), *Rule-governed behavior: Cognition, contingencies, and instructional control* (pp. 191-220). New York: Plenum.

Heagle, A. I., & Rehfeldt, R. A. (2006). Teaching perspective-taking skills to typically developing children through derived relational responding. *Journal of Early and Intensive Behavior Intervention, 3,* 1-34.

Healy, O., Barnes-Holmes, D., & Smeets, P. M. (2000). Derived relational responding as generalized operant behavior. *Journal of the Experimental Analysis of Behavior, 74,* 207-227.

Hofmann, S. G., & Asmundson, G. J. G. (2008). Acceptance and mindfulness-based therapy: New wave or old hat? *Clinical Psychology Review, 28,* 1-16.

Holyoak, K. J., & Thagard, P. (1997). The analogical mind. *American Psychologist, 57,* 35-44.

Horne, P. J., & Lowe, C. F. (1996). On the origins of naming and other symbolic behavior. *Journal of the Experimental Analysis of Behavior, 65,* 185-241, 341-353.

Howlin, P., Baron-Cohen, S., & Hadwin, J. (1999). *Teaching children with autism to mind-read: A practical guide.* Chichester, UK: John Wiley & Sons.

Jacobson, N. S., Dobson, K. S., Truax, P. A., Addis, M. E., Koerner, K., Gollan, J. K., et al. (1996). A component analysis of cognitive-behavioral treatment for depression. *Journal of Consulting and Clinical Psychology, 64,* 295-304.

Jeannerod, M. (1994). The representing brain: Neural correlates of motor intention and imagery. *Behavioral and Brain Sciences, 17,* 187-245.

Ju, W. C., & Hayes, S. C. (2008). Verbal establishing stimuli: Testing the motivative effect of stimuli in a derived relation with consequences. *Psychological Record, 58,* 339-363.

Kanfer, F. H., & Saslow, G. (1969). Behavioral diagnosis. In C. M. Franks (Ed.), *Behavior therapy: Appraisal and status* (pp. 417-444). New York: McGraw-Hill.

Kanter, J. W., Baruch, D. E., & Gaynor, S. T. (2006). Acceptance and commitment therapy and behavioral activation of depression: Description and comparison. *Behavior Analyst, 29,* 161-185.

Kanter, J. W., Busch, A. B., Weeks, C. E., & Landes, S. J. (2008). The nature of clinical depression: Symptoms, syndromes, and behavior analysis. *Behavior Analyst, 31,* 1-21.

Kanter, J. W., Manos, R. C., Busch, A. B., & Rusch, L. C. (2008). Making behavioral activation more behavioral. *Behavior Modification, 32,* 780-803.

Kantor, J. R. (1970). An analysis of the experimental analysis of behavior (TEAB). *Journal of the Experimental Analysis of Behavior, 13,* 101-108.

Karekla, M., Forsyth, J. P., & Kelly, M. M. (2004). Emotional avoidance and panicogenic responding to a biological challenge procedure. *Behavior Therapy, 35,* 725-746.

Kohlenberg, R. J., & Tsai, M. (1991). *Functional analytical psychotherapy: Creating intense and curative therapeutic relationships.* New York: Plenum.

Kohlenberg, R., J., & Tsai, M. (1995). I speak, therefore I am: A behavioral approach to understanding problems of the self. *Behavior Therapist, 18,* 113-116, 124.

Kohut, H. (1971). *The analysis of the self.* New York: International Universities Press.

Kohut, H. (1977). *The restoration of the self.* New York: International Universities Press.

Kokinov, B. N., & Petrov, A. A. (2001). Integrating memory and reasoning in analogy-making: The AMBR model. In D. Gentner, K. J. Holyoak, & B. N. Kokinov (Eds.), *The analogical mind: Perspectives from cognitive science* (pp. 59-124). Cambridge, MA: MIT Press.

Kopp, R. R. (1995). *Metaphor therapy.* New York: Brunner/Mazel.

Kosslyn, S. M., Ganis, G., & Thompson, W. L. (2001). Neural foundations of imagery. *Neuroscience, 2,* 635-642.

Lahav, A., Saltzman, E., & Schlaug, G. (2007). Action representation of sound: Audiomotor recognition network while listening to newly acquired actions. *Journal of Neuroscience, 27,* 308-314.

Leahy, R., & Dowd, T. (Eds.). (2002). *Clinical advances in cognitive psychotherapy: Theory and application.* New York: Springer.

Leary, D. E. (1990). Psyche's muse: The role of metaphor in the history of psychology. In E. E. Leary (Ed.), *Metaphors in the history of psychology* (pp. 1-23). New York: Cambridge University Press.

Leichsenring, F. (2005). Are psychodynamic and psychoanalytic therapies effective? A review of empirical data. *International Journal of Psychoanalysis, 86,* 841-868.

Leigland, S. (2005). Variables of which values are a function. *Behavior Analyst,* 28, 133-142.

Linehan, M. (1993). *Cognitive-behavioral treatment of borderline personality disorder.* New York: Guilford Press.

Linehan, M. (1997). Validation and psychotherapy. In A. C. Bohart & L. S. Greenberg (Eds.), *Empathy reconsidered: New directions in psychotherapy* (pp. 353-392). Washington, DC: American Psychological Association.

Lipkens, R., & Hayes, S. C. (2009). Producing and recognizing analogical relations. *Journal of the Experimental Analysis of Behavior, 91,* 105-126.

Lipkens, R., Hayes, S. C., & Hayes, L. J. (1993). Longitudinal study of the development of derived relations in an infant. *Journal of Experimental Child Psychology, 56,* 201-239.

Longmore, R., & Worrel, M. (2007). Do we need to challenge thoughts in cognitive behavior therapy? *Clinical Psychology Review, 27,* 173-187.

Luciano, C., Gómez, I., & Rodríguez, M. (2007). The role of multiple-exemplar training and naming in establishing derived equivalence in an infant. *Journal of the Experimental Analysis of Behavior, 87,* 349-365.

Luciano, C., Rodríguez, M., Mañas, I., Ruíz, F., Berens, N., & Valdivia-Salas, S. (2009). Acquiring the earliest relational operants: Coordination, difference, opposition, comparison, and hierarchy. In R. A. Rehfeldt & Y. Barnes-Holmes (Eds.), *Derived relational responding applications for learners with autism and other developmental disabilities* (pp. 149-170). Oakland, CA: New Harbinger.

Luciano, C., Rodríguez Valverde, M., & Gutiérrez Martinez, O. (2004). A proposal for synthesizing verbal contexts in experiential avoidance disorder and acceptance and commitment therapy. *International Journal of Psychology and Psychological Therapy, 4,* 377-394.

Luciano, C., Valdivia-Salas, S., Cabello-Luque, F., & Hernández, M. (2009). Developing self-directed rules. In R. A. Rehfeldt & Y. Barnes-Holmes (Eds.),

Derived relational responding applications for learners with autism and other developmental disabilities (pp. 335-352). Oakland, CA: New Harbinger.

Luoma, J. B., Hayes, S. C., & Walser, R. D. (2007). *Learning ACT: An acceptance and commitment therapy skills-training manual for therapists.* Oakland, CA: New Harbinger; Reno, NV: Context Press.

Marks, I., & Dar, R. (2000). Fear reduction by psychotherapies. *British Journal of Psychiatry, 176,* 507-511.

Martell, C. R., Addis, M. E., & Jacobson, N. S. (2001). *Depression in context: Strategies for guided action.* New York: Norton.

Matthews, B. A., Shimoff, E., Catania, A. C., & Sagvolden, T. (1977). Uninstructed human responding: Sensitivity to ratio and interval contingencies. *Journal of the Experimental Analysis of Behavior, 27,* 453-467.

Mays, M. (1990). The use of metaphor in hypnotherapy and psychotherapy. *Individual Psychology, 46,* 423-430.

McCullough, L., Kuhn, N., Andrews, S., Kaplan, A., Wolf, J., & Hurley, C. L. (2003). *Treating affect phobia: A manual for short-term dynamic psychotherapy.* New York: Guilford Press.

McHugh, L., Barnes-Holmes, Y., & Barnes-Holmes, D. (2003). *Training perspective-taking with a 4-year-old child.* Paper presented at the First World Congress on ACT, RFT, and the New Behavioral Psychology, Linköping, Sweden.

McHugh, L., Barnes-Holmes, Y., & Barnes-Holmes, D. (2004). Perspective-taking as relational responding: a developmental profile. *Psychological Record, 54,* 115-144.

McHugh, L., Barnes-Holmes, Y., & Barnes-Holmes, D. (2009). Understanding and training perspective-taking as relational responding. In R. A. Rehfeldt & Y. Barnes-Holmes (Eds.), *Derived relational responding applications for learners with autism and other developmental disabilities* (pp. 281-300). Oakland, CA: New Harbinger.

Michael, J. (1975). Positive and negative reinforcement: A distinction that is no longer necessary; or better ways to talk about bad things. *Behaviorism, 3,* 33-45.

Michael, J. (1986). Repertoire-altering effects of remote contingencies. *Analysis of Verbal Behavior, 4,* 10-18.

Michael, J. (1993). Establishing operations. *Behavior Analyst, 16,* 191-206.

Moore, J. (2008). *Conceptual foundations of radical behaviorism*. Cornwall-on-Hudson, NY: Sloan Publishing.

Morris, E. K., & Todd, J. T. (1998). Watsonian behaviorism. In W. O'Donohue & R. Kitchener (Eds.), *Handbook of behaviorism* (pp. 15-69). San Diego, CA: Academic Press.

Newman, M. G., Castonguay, L. G., Borkovec, T. D., Fisher, A. J., & Nordberg, S. S. (2008). An open trial of integrative therapy for generalized anxiety disorder. *Psychotherapy: Theory, Research, Practice, Training, 45*, 135-147.

O'Donohue, W., & Kitchener, R. (Eds.). (1998). *Handbook of Behaviorism*. San Diego, CA: Academic Press.

Öhman, A. (2002). Automaticity and the amygdala: Nonconscious responses to emotional faces. *Current Directions in Psychological Science, 11*, 62-66.

O'Hora, D., Barnes-Holmes, D., & Roche, B. (2001). Developing a procedure to model the establishment of rule governance. *Experimental Analysis of Human Behavior Bulletin, 19*, 14-16.

O'Hora, D., Barnes-Holmes, D., Roche, B., & Smeets, P. M. (2004). Derived relational networks and control by novel instructions: A possible model of generative verbal responding. *Psychological Record, 54*, 437-460.

Öst, L. G. (2008). Cognitive behavior therapy for anxiety disorders: 40 years of progress. *Nordic Journal of Psychiatry, 62*, 5-10.

Öst, L. G., Thulin, U., & Ramnerö, J. (2004). Cognitive behavior therapy vs. exposure in vivo in the treatment of panic disorder with agrophobia [sic]. *Behaviour Research and Therapy, 42*, 1105-1127.

Overskeid, G. (2008). They should have thought about the consequences: The crisis of cognitivism and a second chance for behavior analysis. *Psychological Record, 58*, 131-151.

Parrott, L. (1984). Listening and understanding. *Behavior Analyst, 7*, 29-39.

Parrott, L. (1987). Rule-governed behavior: An implicit analysis of reference. In S. Modgil & C. Modgil (Eds.), *B. F. Skinner: Consensus and controversy* (pp. 265-276). Philadelphia: Falmer Press.

Peterson, A. L., & Azrin, N. H. (1992). An evaluation of behavioral treatments for Tourette syndrome. *Behaviour Research and Therapy, 30*, 167-174.

Power, M., & Dalgleish, T. (1997). *Cognition and emotion: From order to disorder*. London: Psychology Press.

Ramnerö, J., & Törneke, N. (2008). *The ABCs of human behavior: Behavioral principles for the practicing clinician.* Oakland, CA: New Harbinger; Reno, NV: Context Press.

Rehfeldt, R. A., & Barnes-Holmes, Y. (Eds.). (2009). *Derived relational responding applications for learners with autism and other developmental disabilities.* Oakland, CA: New Harbinger.

Rehfeldt, R. A., Dillen, J. E., Ziomek, M. M., & Kowalchuck, R. (2007). Assessing relational learning deficits in perspective-taking in children with high-functioning autism spectrum disorder. *Psychological Record, 57,* 23-47.

Rehfeldt, R. A., & Hayes, L. J. (1998). The operant-respondent distinction revisited: Toward an understanding of stimulus equivalence. *Psychological Record, 48,* 187-210.

Remington, B., Hastings, R. P., Kovshoff, H., Degli Espinosa, F., Jahr, E., Brown, T., et al. (2007). Early intensive behavioral intervention: Outcomes for children with autism and their parents after two years. *American Journal of Mental Retardation, 112,* 418-438.

Roche, B., & Barnes-Holmes, D. (2003). Behavior analysis and social constructivism: Some points of contact and departure. *Behavior Analyst, 26,* 215-231.

Roche, B., Barnes-Holmes, D., Smeets, P. M., Barnes-Holmes, Y., & McGeady, S. (2000). Contextual control over the derived transformation of discriminative and sexual arousal functions. *Psychological Record, 50,* 267-291.

Roth, A., & Fonagy, P. (1996). *What works for whom? A critical review of psychotherapy research.* New York: Guilford Press.

Safran, J. D., & Muran, J. C. (2000). *Negotiating the therapeutic alliance: A relational treatment guide.* New York: Guilford Press.

Safran, J. D., & Segal, Z. D. (1990). *Interpersonal process in cognitive therapy.* New York: Guilford Press.

Salkovskis, P. (Ed.). (1996). *Frontiers of cognitive therapy: The state of the art and beyond.* New York: Guilford Press.

Sallows, G. O., & Graupner, T. D. (2005). Intensive behavioral treatment for children with autism: Four-year outcome and predictors. *American Journal of Mental Retardation, 110,* 417-438.

Schlinger, H. D. (1990). A reply to behavior analysts writing about rules and rule-governed behavior. *Analysis of Verbal Behavior, 8,* 77-82.

Schlinger, H. D. (2008). Listening is behaving verbally. *Behavior Analyst, 31,* 145-161.

Schnaitter, R. (1978). Private causes. *Behaviorism, 6,* 1-12.

Sidman, M. (1971). Reading and auditory-visual equivalences. *Journal of Speech and Hearing Research, 14,* 5-13.

Sidman, M., & Cresson, O. (1973). Reading and crossmodal transfer of stimulus equivalences in severe retardation. *American Journal of Mental Deficiency, 77,* 515-523.

Sidman, M., & Tailby, W. (1982). Conditional discrimination versus matching to sample: An expansion of the testing paradigm. *Journal of the Experimental Analysis of Behavior, 37,* 5-22.

Siegel, D. J. (1999). *The developing mind: Toward a neurobiology of interpersonal experience.* New York: Guilford Press.

Skinner, B. F. (1938). *Behavior of organisms.* New York: Appleton-Century-Crofts.

Skinner, B. F. (1945). The operational analysis of psychological terms. *Psychological Review, 52,* 270-277.

Skinner, B. F. (1953). *Science and human behavior.* New York: Macmillan.

Skinner, B. F. (1957). *Verbal behavior.* New York: Crofts-Century-Crofts.

Skinner, B. F. (1963). Behaviorism at fifty. *Science, 140,* 951-958.

Skinner, B. F. (1966). An operant analysis of problem solving. In B. Kleinmuntz (Ed.), *Problem solving: Research, method, and theory* (pp. 133-171). New York: John Wiley & Sons.

Skinner, B. F. (1974). *About behaviorism.* New York: Knopf.

Skinner, B. F. (1989). *Recent issues in the analysis of behavior.* Columbus, OH: Merrill Publishing.

Smith, L. J. (1996). A behavioural approach to the treatment of non-retentive encopresis in adults with learning disabilities. *Journal of Intellectual Disability Research, 40,* 130-139.

Solso, R. L., MacLin, M. L., & MacLin, O. H. (2005). *Cognitive psychology* (7th ed.). Boston: Pearson Education International.

Stewart, I., & Barnes-Holmes, D. (2001). Understanding metaphor: A relational frame perspective. *Behavior Analyst, 24,* 191-199.

Stewart, I., & Barnes-Holmes, D. (2008). *A functional-analytic model of metaphor using the relational evaluation procedure.* Unpublished manuscript.

Stewart, I., Barnes-Holmes, D., & Roche, B. (2004). A functional-analytic model of analogy using the relational evaluation procedure. *Psychological Record, 54*, 531-552.

Stewart, I., Barnes-Holmes, D., Roche, B., & Smeets, P. M. (2001). Generating derived relational networks via the abstraction of common physical properties: A possible model of analogical reasoning. *Psychological Record, 51*, 381-408.

Stewart, I., Barnes-Holmes, D., Roche, B., & Smeets, P. M. (2002). A functional-analytic model of analogy: A relational frame analysis. *Journal of the Experimental Analysis of Behavior, 78*, 375-396.

Stewart, I., & McElwee, J. (2009). Relational responding and conditional discrimination procedures: An apparent inconsistency and clarification. *Behavior Analyst, 32*, 309-317.

Strosahl, K. D., Hayes, S. C., Wilson, K. G., & Gifford, E. V. (2004). An ACT primer. Core processes, intervention strategies, and therapist competencies. In S. C. Hayes & K. D. Strosahl (Eds.), *A practical guide to acceptance and commitment therapy* (pp. 31-58). New York: Springer-Verlag.

Tomkins, S. (1987). Script theory. In J. Aronoff, A. I. Rabin, & R. A. Zucker (Eds.), *The emergence of personality* (pp. 147-216). New York: Springer.

Törneke, N., Luciano, C., & Valdivia-Salas, S. (2008). Rule-governed behavior and psychological problems. *International Journal of Psychology and Psychological Therapy, 8*, 141-156.

Tsai, M., Kohlenberg, R. F., Kanter, J. W., Kohlenberg, B., Follette, W. C., & Callaghan, G. M. (2008). *A guide to functional analytic psychotherapy: Awareness, courage, love, and behaviorism*. New York: Springer.

Valdivia, S., Luciano, C., & Molina, F. J. (2006). Verbal regulation of motivational states. *Psychological Record, 56*, 577-595.

Vosniadou, S., & Ortony, A. (1989). *Similarity and analogical reasoning*. Cambridge, UK: Cambridge University Press.

Vygotsky, L. (1986). *Thought and language*. Cambridge, MA: MIT Press.

Watson, J. B. (1929). *Psychology from the standpoint of the behaviorist* (3rd ed.). Philadelphia: Lippincott.

Wegner, D. M., & Gold, D. B. (1995). Fanning old flames: Emotional and cognitive effects of suppressing thoughts of a past relationship. *Journal of Personality and Social Psychology, 68*, 782-792.

Whelan, R., & Barnes-Holmes, D. (2004a). The transformation of consequential functions in accordance with the relational frames of same and opposite. *Journal of the Experimental Analysis of Behavior, 82*, 177-195.

Whelan, R., & Barnes-Holmes, D. (2004b). Empirical models of formative augmenting in accordance with the relations of same, opposite, more-than, and less-than. *International Journal of Psychology and Psychological Therapy, 4*, 285-302.

Whelan, R., Barnes-Holmes, D., & Dymond, S. (2006). The transformation of consequential functions in accordance with the relational frames of more-than and less-than. *Journal of the Experimental Analysis of Behavior, 86*, 317-335.

Williams, J. M. G. (1992). *The psychological treatment of depression.* London: Routledge.

Wilson, K. G. (2001). Some notes on theoretical constructs: Types and validation from a contextual behavioral perspective. *International Journal of Psychology and Psychological Therapy, 1*, 205-215.

Wilson, K. G., O'Donohue, W. T., & Hayes, S. C. (2001). Hume's psychology, learning theory, and the problem of knowledge amplification. *New Ideas in Psychology, 19*, 1-25.

Wray, A. M., Freund, R. A., & Dougher, M. J. (2009). A behavior-analytic account of cognitive bias in clinical populations. *Behavior Analyst, 32*, 29-49.

Young, J. E. (1990). *Cognitive therapy for personality disorders: A schema-focused approach.* Sarasota, FL: Professional Resource Press.

Young, J. E., Klosko, J. S., & Weishaar, M. E. (2003). *Schema therapy: A practitioner's guide.* New York: Guilford Press.

Zettle, R. D., & Hayes, S. C. (1982). Rule-governed behavior: A potential theoretical framework for cognitive behavior therapy. In P. C. Kendall (Ed.), *Advances in cognitive behavioral research and therapy* (pp. 73-118). New York: Academic Press.

Niklas Törneke, MD, is a psychiatrist and licensed psychotherapist in private practice in Kalmar, Sweden. Together with Jonas Ramnerö, Ph.D, he has previously authored *The ABCs of Human Behavior.*

Foreword writer **Dermot Barnes-Holmes, Ph.D.**, is foundation professor of psychology at the National University of Ireland, Maynooth, and coauthor of *Relational Frame Theory.* He has published over two hundred scientific articles, and book chapters, the vast majority of which have focused on the study of human language and cognition from a behavior-analytic perspective.

Afterword writer **Steven C. Hayes, Ph.D.**, is University of Nevada Foundation Professor of Psychology at the University of Nevada, Reno, and has authored and coauthored numerous books, including *Acceptance and Commitment Therapy, Relational Frame Theory* and *Get Out of Your Mind and Into Your Life,* as well as hundreds of articles and book chapters on related subjects.

Index

C

D

E

F

P

R

S